Breakfast of Biodiversity

The Truth about Rain Forest Destruction

John Vandermeer
and
Ivette Perfecto

Foreword by Vandana Shiva

A Food First Book
The Institute for Food and Development Policy
Oakland, California

Managing Editor: Kathleen McClung
Project Editor: Michael Z. Jody
Proofreader: Theo Crawford
Cover Designer: Lory Poulson
Design & Typesetting: Harvest Graphics

Library of Congress Cataloging-in-Publication Data

Vandermeer, John H.
      Breakfast of biodiversity : the truth about rain forest destruction / by
John Vandermeer and Ivette Perfecto ; edited by Michael Z. Jody.
      p.   cm.
      "A Food first book."
      Includes bibliographical references (p.   ).
      ISBN 0-935028-66-8 : $16.95
      1. Rain forest conservation.   2. Rain forest ecology.   3. Banana
trade—Environmental aspects—Costa Rica.   4. Deforestation-
Environmental aspects—Tropics.   5. Agricultural ecology—Tropics.
6. Biological diversity conservation—Tropics.   I. Perfecto, Ivette.   II. Jody,
Michael Z.   III. Title.
SD414.T76V36   1995
333.75'137'0913—dc20   95-6598

Printed in USA
10 9 8 7 6 5 4 3
To order additional copies, please write or call our distributor:
Subterranean Company
Box 160, 265 South Fifth Street
Monroe, OR   97456
800-274-7826

Dedication

This book is dedicated to Tina and Kiko,
and to the poor farmers of the tropics.

Table of Contents

Illustrations

Tables

Foreword

by Vandana Shiva

John Vandermeer and Ivette Perfecto have in their book *Breakfast of Biodiversity* raised some of the most challenging questions about conservation of biodiversity.

The old conflict of ecology versus economics, environment versus development, has, in the post Rio era, mutated into a conflict about how natural resources will be conserved, by whom and for whom.

The dominant paradigm sees conservation as dependent on financial investments, which are, in turn, linked to increased economic growth, international trade and consumption. This approach allows continued destruction of the environment and people's livelihoods in the domain of the productive economy, while allowing islands of "set-asides" and "wilderness" reserves, which also displace people and destroy indigenous cultures and lifestyles. In this paradigm, there is an artificial separation of conservation and production, of people and nature. Further, the increased trade and commerce that generate the financial resources central to this paradigm of conservation themselves destroy natural resources, biodiversity, cultural diversity and people's livelihoods. The result is nonsustainable islands of biodiversity threatened with constant erosion by a sea of pollutants and monocultures.

The second paradigm is based on conserving biodiversity as the very basis of production, which ensures that both nature and people's livelihoods are protected. It is this second paradigm that is both people- and nature-friendly that this book articulates. In this paradigm, conservation cannot be isolated from production, investment and trade. Environmental action becomes inseparable from issues of social justice and peoples' economic rights. As the authors state in closing, "calls for boycotts of tropical timbers or bananas need to be coupled with actions to change investment patterns and international banking pressures."

Preface

The world is already saturated with books about rain forests. What then can be our excuse for writing yet another? It's simple. After reading many of those other books in the course of preparing slide shows, lectures and discussions, we came to believe that for the most part they really did not get it completely right. While they all made very important points about the nature of the rain forest and the alarming rate of its disappearance, and they all had particular analyses about causes, their analysis focused on one or another issue — overpopulation, export agriculture, peasant agriculture, etc.

We can appreciate the temptation to focus on the facts of rain forest destruction, and we agree that the nature of the problem itself is quite worthy of persistent propaganda. This, we suppose, is why all the books say the same thing — tropical rain forests are useful and beautiful, yet they are being destroyed. That the problem needs to be brought to the attention of the public, we agree. But once alerted to the problem, the public asks what to do. Causes must be addressed, and we feel that most of the popular literature on the subject does not do it adequately.

To be sure, many books talk about causes. But most frequently, authors are concerned with identifying some "ultimate factor" — overpopulation, greedy lumber companies, inefficient peasant farmers, avaricious export agriculture. We agree that some, even many, of these forces are part of the picture. But in the final analysis the cause is far more complicated. Indeed, the nature of the complications *is* the cause. This sort of analysis is multifaceted, with many interconnecting components — what we refer to as the "web of causality." However, we also feel that it is not difficult to appreciate this analysis, if the focus is not on an individual component but rather on the complete web. That is the purpose of this book.

We aim to alert the reader to several obvious facts. The web includes subjects that ordinarily do not occur together between the same book covers: the poor soils of peasant farmers, international diplomacy, international agricultural economics, and a variety of other strands in the web. Thus our narrative ranges from the acidity of rain forest soils to the acridity of international politics.

Furthermore, this book is in no sense complete. It is not a comprehensive analysis of the nature of rain forests and their potential utility, nor is it a full presentation of the details of rain forest destruction the world over. Many other books do precisely that, and they do it quite well. Our purpose is to elaborate, in a compact and straightforward manner, the complicated story of why rain forests are disappearing. This is mainly a social, economic, and political story, with a pinch of ecology. The story has nothing to do with overpopulation and is not about a few evil capitalists who care more for profits than trees. Rain forests thus will not be saved by handing out condoms nor by refusing to buy furniture made of tropical wood. The only way to reverse the pattern of the past five hundred years is, first, to understand the complexity of the web that creates the problem in the first place and, second, to develop a strategy that shreds that causal web. Perhaps a boycott of tropical woods would make sense, but only within the context of a clear analysis of how that boycott contributes to eliminating the web of causality.

While our purpose is to elaborate the web of destruction in a brief form, we frequently find it easiest to argue from example. Since our personal experience is mainly in the tropical rain forests of Central America, almost all of our examples are from there. Other authors would obviously have used different examples. But in the end, our analysis is not restricted to the Central American case. To be sure, there will be differences from place to place — logging is more important in southeast Asia, for example, cattle ranching more important in Brazil. But the general principles, and consequently the appropriate actions, are not dependent on the particular site. The web is strung slightly differently in each place, but its connecting strands are similar the world over.

It would be extremely difficult for us to remember the many individuals who have contributed to our formulation of this issue. For the most part they know who they are, and we thank them. In particular, we wish to thank the "Bluefields group" at the University of Michigan; the professors and students in the Department of Ecology and Natural Resources at the Universidad Centroamericana, Nicaragua; several farmers from El Progreso, Costa Rica; the members of La Fonseca and La Union cooperatives near Bluefields, Nicaragua; and finally several Mexican farmers who alerted us to a more sophisticated vision of the aesthetics of rain forests than we had seen earlier. Special thanks are due to Victor Cartín, Marianos Azofefa, Ernesto Lemos, and Doug Boucher. Mike Jody patiently untangled much of our prose, for which the reader should be just as grateful as we are.

John Vandermeer and Ivette Perfecto
Ann Arbor, Michigan
February, 1995

1

———◆———

Slicing up the Rain Forest
on Your Breakfast Cereal

The eastern slopes of the Barva volcano catch water-laden trade winds from the Caribbean to create the climate of Costa Rica's eastern coast, location of some of the most beautiful rainforests in the world. Here you can experience that special feeling that inspires poets and explorers — from the myriad vegetative forms so evident even on first glance (figure 1.1) to the misty mornings that invoke mysterious feelings and bucolic images of paradise lost. The rain forest here, as elsewhere, is a collective human construct that sometimes serves as our mystical Garden of Eden, but is also a material collection of fabulous plants and animals, a natural construct of the high temperature and heavy rainfall of equatorial climates. The trade winds rise as they encounter the eastern seaboard and with their ascent they cool, condensing the water vapor they borrowed during their voyage across the Caribbean. The consequent rains collect in several basins and come together roughly at the town of Puerto Viejo, continuing northward to empty into the San Juan River, the border with Nicaragua. Thus is the region known as the *Sarapiquí* (sah rah pick ee, with an accent on the ee),

Figure 1.1. The lowland tropical rain forest of Central America. Note the wide variety of vegetative forms evident.

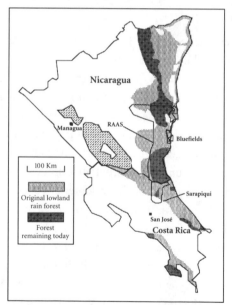

Figure 1.2. Map of Nicaragua and Costa Rica, illustrating the original and current extent of lowland tropical rain forest, and the position of the two political units, the *Sarapiquí* county of Costa Rica and the RAAS (*Región Autonoma del Atlántico Sur*) of Nicaragua (further discussed in Chapter 7).

site of several of the world's most famous rainforest conservation areas (figure 1.2).

Streaming into the area to partake of the breathtaking beauty of the natural world in this "one of the last" pristine places in the world, are biologists, ecologists, and ecotourists, spending their grant money or retirement savings to visit the "heritage of humanity." It is hardly necessary to repeat the cliches any longer — tropical rainforests cover only 7% of the earth's surface yet harbor at least 50% of the world's plant and animal species (the earth's biodiversity); they are the lungs of the world, eating away at the excessive carbon dioxide we have excreted from our industrial metabolism; they are the source of foods and pharmaceuticals, bananas and Brazil nuts, chocolate, cashews, coffee and cocaine, cortisone and quinine. They are also beautiful! The aesthetics of these forested lands cannot be overestimated, and the sense of wonder one experiences walking through this cradle of biodiversity cannot be expressed in words.

But, as anyone visiting the *Sarapiquí* can readily see, all is not well in this Garden of Eden. Certainly, it remains beautiful inside of the conservation areas. The problem is outside those areas. And the problem is the same one Costa Rica has had ever since Minor C. Keith built his famous railroad, and helped found the United Fruit Company in 1898. The problem is the banana. Currently at least five major banana companies are converting vast acreage in the area to banana plantations, thus threatening both directly and indirectly the rain forests we so revere. Those same biologists, ecologists, and ecotourists, who love the rain forest when they're in Costa Rica, also love to slice bananas onto their cereal in the morning. And with our penchant for viewing the world in isolated little disconnected fragments, it is apparently difficult for us all to see the connection between the knife that slices the banana into our cereal bowl, and the chain saw that slices tree trunks onto the rain forest floor.

Not so long ago, environmental activists in the developed world became aware of the so-called "hamburger connection." Central American rain forests were being cut down at an alarming rate to make way for the production of low quality beef to supply the fast-food industry in the First World. Stop eating fast-food hamburgers, the argument went, and you would reduce the demand to cut down more forest. The banana expansion currently underway in Central America has been likened to this hamburger connection. But the whole argument surrounding the hamburger connection was flawed, and an attempt to construct the same argument for bananas would simply repeat that flaw. In fact, the expansion of bananas, like the expansion of pasture for beef production, is a tangled web of subtle connections. Tweak the web at one point and it reverberates all over, sometimes in unexpected ways. Understanding the nature of the connections in this "web of causality" is the purpose of this book.

The transformation currently underway in the *Sarapiquí* is neither unprecedented nor unique, which makes it a useful example. Similar patterns occur throughout the tropics. Sometimes the pattern involves bananas, sometimes cattle, maybe citrus, African oil palm, or rubber trees — a variety of commodities, similar politically if quite distinct biologically. The pattern is a six stage process: 1) Visionary capitalists identify an economic opportunity for the market expansion of an agricultural product. In this case, the opportunity is the opening up of markets in Eastern Europe and the unification of Western Europe, and the product is bananas. 2) They purchase (or steal, or bribe into a government concession) some land, including land that may contain rain forest, which is promptly cut down. 3) They import workers to produce the product (in this case workers come from all over Costa Rica and even from Nicaragua). 4) After a period of boom the product goes bust on the world market, which means scaling back production, which in turn means releasing a significant fraction of the work force. 5) The newly unemployed work force seeks and fails to find employment elsewhere, and must seek land to grow subsistence crops to tide themselves over until other work can be found. And finally, 6) the only place the now unemployed workers can find land no one will kick them off of is in the forest, which means yet more of the rain forest is converted to agriculture.

In this way Costa Rica, one of the world's showcases of conservation, is currently promoting a policy that actually encourages rain forest destruction. That is interesting by itself, but this specific example is not as important as the general idea it highlights. The crop in this instance happens to be bananas, but the general pattern is all too common.

Costa Ricans and Their Love/Hate of Bananas

An afternoon in Puerto Viejo, the little town located near the con-fluence of the rivers draining the Barva volcano, reveals what might surprise a European or North American tourist. Despite the fact that, given their history, Costa Ricans understandably love to hate bananas, it is difficult to find anyone in town who does not fully support the massive banana expansion currently underway. Furthermore, the gov-ernment, both local and national, is encouraging the expansion with a vigor normally associated with a depressed Northern U.S. city courting a big assembly plant. (*You want no unions? You got it. You want tax breaks? Just say how much. You want license to pollute? Smoke your heart out. But please, just locate here.*) Costa Rica is as debt-laden as the rest of Latin America,[1] and needs all the money it can get just to service its debt. The expansion of bananas is one way to make money. Thus, despite the recognition that social and environmental problems will inevitably come along with the bananas, the vast majority of Costa Ricans, both in the *Sarapiquí* and elsewhere, welcome the current expansion. A small group of Costa Rican environmentalists are pro-testing, but they are overwhelmed by more powerful interests.

Of the approximately quarter of a million hectares[2] in the *Sarapiquí* valley, some 50,000 are devoted to biological preserves.[3] Another 100,000 hectares are in the legal hands of small peasant farming com-munities. The rest (approximately 100,000 hectares) is a mosaic of small farms, most without title to the land; secondary and old-growth forest; cattle pasture; and an occasional sizable ornamental-plant or fruit plantation.[4]

In the periphery of the valley lies an extensive banana plantation owned by the Standard Fruit Company, a major employer in the region for the past quarter century. The history of Standard Fruit provides an example of what might be expected on a larger scale in the near future. Tales are common of pesticide abuse, waste-dumping into local waterways, deforestation, and the massive social problems normally associated with a frontier area. Best documented is the celebrated case in which Standard Fruit was accused of negligence in its use of DBCP (dibromochloropropane), a popular fungicide.[5] During the early 1970's more than 2,000 banana workers were rendered sterile by this poison. They are currently suing Standard Fruit in a United States court. This and other past records indicate that historically the banana companies have not accepted responsibility for the health and safety of their workers, the community, or the environment. With the current massive banana expansion there is no reason to assume that these adverse environmental, social, and health effects will not be repeated on an even larger scale.

History is perhaps even more ominous when we examine another long-term pattern evidenced by the Standard Fruit operation in the area. Standard Fruit employs workers who migrate to the *Sarapiquí* from other parts of Costa Rica. These workers are retained as long as the market for bananas is sufficiently robust, but are let go when sales slacken. The laid-off workers are mainly rural people, former peasants drawn into rural-wage labor. In past decades the ebb and flow of the banana business has created critical periods in which many workers were laid off and forced to fend for themselves. These layoffs were a natural product of the world economic system, due both to fluctuating banana prices, and to the very existence of a two-part economy — export bananas on the one hand and worker/farmer on the other. There is little employment opportunity in the area other than the banana companies, so when workers are laid off they must either migrate to the city, adding to the growing shanty towns, or they must turn to farming. In order to farm they have to find a homestead. Sometimes that small piece of land is in a rain forest. Other times it is a small corner of some large absentee landowner's cattle ranch, in which case, depending on complicated criteria, either the homesteading family is eventually forcibly evicted, or the state agrarian reform institute adjudicates a "fair" purchase for the peasant family.

The past thirty or forty years have seen this arrangement persist, with rain forest cover in the region plummeting from almost 90% in 1950 to approximately 25% today. Only a small portion of the remaining 25% is not part of one of the four large biological reserves.

Loggers, Farmers, and Banana Companies — a Rich History[6]

This pattern, so readily observable today, is set in a rich ecological history beginning well before the current crisis. Early in the century, extensive river systems were used to transport both logs and bananas. Logging was a rather small-scale operation by modern standards, but had the effect of drawing workers to the area and creating pathways into the forest. Since only a handful of the many species of trees in the rain forest were actually valuable, it was necessary to scout out and then cut a path to the valuable trees, and after cutting them down to haul them out with teams of oxen or horses. By the 1930's, the land along almost all the rivers was deforested and planted with bananas, while the surrounding forest was riddled with trails made for dragging logs. The logging process intensified in the late 1940's and 50's when machinery was brought into the area, and a complex network of logging roads crisscrossed what forests remained after the inroads already made by the banana plantations. Men who originally came to the area to work in the logging industry used these roads to gain access to logged areas and frequently established homesteads. Former banana workers did the same thing.

In the late 1940's everything changed in the *Sarapiquí*, as it did throughout the Atlantic coast of Costa Rica. Devastating fungal diseases routed the banana industry. The extensive plantations of the United Fruit Company and a variety of independent producers were decimated by this disease. No cure could be found and the company moved its entire operation to the other side of the mountains, where the disease had not been established. It was not until the mid 1950's that a genetic variety of banana that was resistant to the disease was developed, thus enabling the Standard Fruit Company to establish its plantations in the area in the late 1950's.

Bananas Today

Now the plot is thickening. In anticipation of an expected surge in the demand for bananas (the anticipated result of opening markets in Eastern Europe, and the economic unification of Western Europe), five or six major banana companies have been purchasing large expanses of land and expanding banana production accordingly. The area planted to banana rose from 20,000 hectares in 1985 to 32,000 hectares in 1991, and visits to the area in 1991 and 1992 revealed intense activity in setting up new banana plantations throughout the valley. As much as 45,000 hectares are expected to be in bananas by the end of 1995.[7] As had happened in the past, workers were drawn from all over the country. But breaking with past traditions, this time there apparently were not enough Costa Rican workers to do the necessary work, and workers were also attracted from Nicaragua, Panama, and even Honduras. It appears likely that within two years most if not all of the arable land not currently in either biological preserve or organized peasant agricultural communities, will become banana plantations.

A variety of factors make Costa Rica, and particularly the *Sarapiquí* basin, an especially attractive target of the banana companies. Notably, the infamous *Solidarista* movement has destroyed all union activity in the area. Some ten years ago, this church-based, U.S. supported, anti-union movement systematically moved into the *Sarapiquí* valley to replace all banana labor unions with a new concept for worker organization. *Solidarista* dogma outlaws strikes, does not recognize the right of workers to collectively bargain, and seeks to attract workers with frivolous benefits such as clubhouses and soccer fields. With massive funding from the Association for Free Labor Development, the international wing of the AFL-CIO long suspected to have CIA ties, democratic labor unions were systematically attacked throughout Costa Rica. The campaign was especially effective in the *Sarapiquí* where union membership now stands at zero and company officials proudly proclaim that no union people are able to find jobs.[8] A local political official told us in 1991 that the planned banana expansion would have been impossible without the existence of the *Solidarista* organizations.

A second important factor is the willingness of the Costa Rican government and its partner, the United States of America, to create infrastructural conditions, which favor the banana companies. Roads are being constructed, bridges are being built, hospitals and schools are planned, all for the purpose of creating an attractive infrastructure for the banana companies. The U.S. Army Corps of Engineers was enlisted in this effort. In a 1992 program called "Bridges for Peace," Army Corps engineers built roads and bridges in the *Sarapiquí*. A cynical U.S. serviceman told us the program has been unofficially dubbed "Bridges for Bananas," as the construction so obviously focuses on improving infrastructure for the export of bananas. U.S. Army engineers built many of the roads and bridges that today carry the logs of the cut rain forest, and tomorrow will carry the harvest of the international banana companies. Indeed, with the infrastructure provided by U.S. taxpayers at the request of the Costa Rican government, from roads and bridges to the *Solidaristas*, from the "converted" rain forest to new social infrastructure, investment opportunities look good — that is, if you are a banana company.

But the banana companies, mindful that their operations might attract outside attention, were prepared to pay "expert" scientists to mollify the public. Corporación Bananera Nacional (CORBANA), was formed in 1990. Some twenty years earlier a smaller national effort, Asociación Bananera Nacional (ASBANA) had been formed by the Costa Rican government for the purpose of developing technical assistance for small producers of bananas in the country. Operating on a tiny budget, this small research operation persisted until two years ago, when the rush to privatization caught up with it and ASBANA changed to CORBANA and began to receive money directly from the banana companies. For every box of bananas exported, each company pays a fee to support the research efforts of CORBANA. Theoretically CORBANA conducts research aimed at making banana production more environmentally friendly. This research was to include proposed projects on using biotechnology to develop strains of bananas resistant to pests, development of organic fertilizers, and extensive surveys of fauna in the banana plantations (ostensibly to monitor the effects of the plantation on wildlife). However, in a visit to the CORBANA facility we observed very impressive projects on soils, plant diseases, parasites, and drainage, but none of the celebrated studies to promote environmental friendliness.

But we repeat and emphasize that the expansion of bananas is viewed as a positive event by nearly everyone in the *Sarapiquí*, and by most observers in the entire country. Local workers and peasants see jobs being created, local merchants see a potential surge in business, and local politicians see an increase in their power base. The Costa Rican government itself promotes the expansion since it sees the increased tax revenues as helping to pay debt service on its tremendous international debt. The accepted fact that almost all profit from banana farming will leave the country seems of little local concern. This is perhaps understandable given the state of the local and national economy. But less comprehensible is that segments of the international "conservation" community have come on board, and either retain a calculated obliviousness to what is going on, or actively pursue a neutral position. Significant, yet weak, opposition is coming from a small, loosely-structured local conservation movement, composed of Costa Ricans, and organized without the help of the international conservation community. They are fighting what appear to be insurmountable odds.

Bananas Tomorrow

It is not difficult to predict what is likely to happen next. Unless history has significantly reversed free market laws, banana prices will fluctuate on the world market as they always have. Furthermore we can expect the banana companies to do what they are supposed to as good managers: reduce the cost of labor, which they will do at points of economic downturn by laying off workers, just as they always have in the past. But this time where will those former workers go? In the past there was always that mosaic of small farms, large cattle pastures, second growth, and primary forest between the organized agricultural settlements and the biological preserves. Such areas were normally considered

(aid: no halters when banana prices fluctuate)

externalities, sopping up the overflow of humanity that could not be accommodated in times of crisis. Now, however, that area will be taken up by banana plantations, and the only remaining area not already devoted to some form of agriculture (and therefore "available"), will be within the four biological preserves in the area. It would take enormous naïveté to suppose that when their survival is at stake, these landless peasants will not begin cutting forest in the biological preserves.

This example illustrates the dynamics that occur, with different details of course, throughout Central America and in much of the rest of the world. Because of the nature of the world economic system, Costa Rica really has no choice but to promote the expansion of bananas. Costa Rica's international debt, accumulated because of its position in the world economy, and its need for the expansion of international capital, require that it seek tax revenues however it can. The banana companies themselves (at least one of which is Costa Rican) continue to play their historical role as international accumulators of capital, and temporary employers of peasants, thus maintaining the dysfunctional two-part economy. Peasants continue arriving from other parts of the country and now, even from Nicaragua, seeking jobs and the "good life" and willing to accept minimal conditions — but since the *Solidarista* movement *(laborers)* destroyed the unions, they are without significant political representation. The stage was initially set by the loggers with their systems of logging roads, and the first wave of banana plantations with their periodic layoffs that forced peasants into the forests. If the process continues with this basic overall structure, which we see no reason to doubt, there is little hope in the long run, even for those rain forests under protected status.

Current rumors in the area indicate that the long-running Standard Fruit Company operation in Rio Frio, will soon abandon all of their plantations near the *Sarapiquí.* Despite what some promoters claim, banana plantations do not last forever. A variety of ecological forces eventually catch up with such intensive production, and the plantations must be abandoned. Will the legacy of bananas leave the area with much degraded conditions of production, as has happened repeatedly in the past? Who will bear the costs of restoring conditions to their original state, if that will even be possible? And who will share concern with the thousands of rural people, deprived of their land and their livelihood with no place to go? Who will tell them that the rain forest is more important than feeding their families?

Viewing the Problem from Various Perspectives

In the midst of these dramatic events, an internationally recognized ecologist gave a public lecture at a local ecotourism center in the *Sarapiquí*, claiming that recent deforestation in the area was due to the inevitable march of Malthusian reality. He claimed that overpopulation was causing the destruction of the forest. In a sense, of course, such an observation is trivially true. There undoubtedly is an overpopulation of banana companies, an overpopulation of former banana workers looking for land, an overpopulation of adventurers seeking their fortunes in a new frontier zone, an overpopulation of greedy people and institutions, and even an overpopulation of ecotourists from Europe and the United States.

But, when this expert ecologist declared that a Malthusian crunch was the root of the problem, he was actually implying something rather different — that the pressure of having too many children, the birth rate of the population, is the real problem. This point of view implies that the main solution to the problem is birth control. It further implies that this is a sufficient solution, that it is useless to do anything other than promote birth control, and that as long as population densities remain as they are, the pattern of deforestation will continue.

An alternative viewpoint, expressed to us by a local conservationist, is that avaricious banana companies are cutting down rain forests because they are hungry for profits. They will stop at nothing to satisfy their need to accumulate ever greater quantities of capital and the forests will continue to disappear as long as banana companies are allowed to continue their greedy operations. This also is a distinct point of view. It implies that the only solution to the problem is to eliminate the capitalist. It further implies that this is a sufficient solution, that it is useless to do any-

[handwritten margin note: solutions 1) eliminate over-pop. 2) eliminate capitalist]

thing other than "smash capitalism," and that as long as the need to accumulate capital remains, the pattern of deforestation will continue.

These points of view are prisms through which the facts of the matter may be interpreted. They both encourage a single focus for solution — stop population growth, or smash capitalism. We believe that both are right in a very limited sense. But we also believe that they are both wrong in a broader, and more practical sense. Ultimately each of these prisms focuses on a single thread in a fabric of causality. Eliminating one thread will not eliminate the problem. The problem is the fabric itself. The proper means of understanding the situation then, is to look at the complicated way that various forces are interdependent, especially focusing on the way countervailing tendencies are resolved. The approach we take in this book may, at first glance, seem as narrow as the approaches of those who advocate population reduction, or smashing capitalism. We assert that food insecurity is the root cause of deforestation. It is a critical thread in the fabric of causality. We take this approach for two reasons. First, we wish to provide an antidote to the simplistic views that either overpopulation, or avaricious capitalism, cause deforestation. Second we will argue that, given the ultimate goal of reworking the entire fabric of causality, the place to begin that process is with food security. We will not argue that providing peasants with food security will stop deforestation *per se*, but rather that beginning the political process of reorganizing socioeconomic-ecological systems by examining questions of food security, will force both analysis *and* practice into the realms ultimately necessary for the resolution of this issue. When neo-Malthusians suggest there are too many people for the land base, the food security position reveals several important particulars: that peasants seek land to feed their families, not because there are too many of them or too little land (at least right now), but because available land is occupied by other activities. Our orientation will also reveal that the techniques for sustainable agriculture in that zone have been replaced with destructive, chemically-based ones, and further, that the legal status of most peasants is "landless" even when they clearly occupy a piece of land. When radicals purport that avaricious capitalism causes deforestation, the food security position shows that the evolution of modern agriculture has created international structures that force even progressive governments like Cuba to invite those greedy capitalists into their economies. The international order, which causes food insecurity in the developing world, is implicated in a chain of events that ultimately leads to the transformation from workers to peasants who must seek out rain forest land to farm in order to provide food for their families.

We do not wish to leave the impression, however, that food insecurity is just another mechanistic cause to which the problem of rain forest destruction may be reduced. It is clearly not. But, as a mode of analysis, examining food insecurity will cause us to deal with the entire complex web of ecological, sociological, economic, and political issues on which the poisonous spider of rain forest destruction crouches.

Two Models for Saving the Rain Forests

Current events in the *Sarapiquí* region are, alas, not unique. Tropical rain forest areas around the globe are experiencing similar complex socioeconomic forces that threaten to continue or even accelerate the destruction of this most diverse of all ecosystems. In all of these areas there has been some reaction from local and international concerns. Unfortunately, much of this response is misdirected because it is based on a distorted image of the facts, and on an implicit ideology — what we call the Mainstream Environmental Movement Approach — which allows only a narrow range of possible courses. We feel there is an alternative philosophical approach, the Political Ecology Strategy, which emphasizes basic issues of security: security of land ownership and the consequent ability to produce food for local consumption.

The mainstream environmental movement has raised large sums of money to purchase and protect islands of rain forest with little concern for what happens between those islands, either to the natural world or to the social world of the people who live there. We doubt this strategy has much chance of succeeding. It is likely that in the short term the landscape will be converted into isolated islands of tropical rain forest, surrounded by a sea of pesticide-drenched modern agriculture, underpaid rural workers, and masses of landless peasants looking for some way to support their families. The long term prospects, however, are worse, as the example of the *Sarapiquí* suggests.

Our alternative, the Political Ecology Strategy, emphasizes the land and people *between* the islands of protected forest, and has greater credibility because of its willingness to see some of the interconnections in this complicated system. This point of view has been variously known as *ecological development, sustainable development,* or *eco-development,* though all of these terms have been cynically adopted by even the most environmentally destructive agencies. Whatever we call it, this approach views the problem, properly, as a landscape problem with forests, forestry, agroforestry, and agriculture as interrelated land-use systems, and seeks to develop those land-use systems so that conditions of production according to the needs of the local population may be maintained. The Political Ecology Strategy challenges non-sustainable development projects, such as modern banana plantations, and seeks to organize people to oppose ecologically and socially damaging development.

These two points of view lead to quite different projections of what the future rain forest areas of Central America might look like. If the mainstream position remains dominant, we expect to see, in the short term, a sea of devastation with islands of pristine rain forest, and in the long term, nothing but a sea of devastation. The political ecology point of view envisions a mosaic of land-use patterns: some protected natural forest, some extractive reserve, some sustainable timber harvest, some agroforestry, some sustainable agriculture, and, of course, human settlements. This mosaic would be sustainable over time.

But is this a practical vision in the real world? The decision to promote bananas in the *Sarapiquí* can hardly be faulted on "modern" economic grounds. Sadly, however, if national and international commitment to the archaic economics of Adam Smith and the IMF persists, we fear continuing destruction of the rain forests and the deterioration of the lives of the people of *Sarapiquí.* The alternative requires a radical rethinking of what sorts of economic and political arrangements are to be tolerated, the sort of rethinking that can get you in trouble in Central

America, the sort of rethinking that may even challenge the idea that it
is our inalienable right to slice bananas onto our breakfast cereal.

Costa Rica, Bananas, and a General Pattern

The case study elaborated in this chapter is typical. Granted, there are
cases in which rain forests are being cut with a profoundly different logic
(several areas in South East Asia and much of the Amazon, for example),
but both historical and contemporary patterns the world over reflect the
basic paradigms seen in this example. The details vary, but the underlying logic is remarkably consistent.

Costa Rica has been held up as one of the world's best examples of
rain forest conservation. Its internationally recognized conservation
ethic, its position of relative affluence, its democratic traditions, the
remarkable importance of ecotourism to its national economy, its willingness to adopt virtually any and all programs of conservation promoted by western experts, make it the most likely place for the success
of the traditional model of rain forest conservation. The fact that the
model has been an utter failure in Costa Rica, where it had the greatest
chance of success, calls the model itself into serious question.

This case study is intended as an outline of the general problem. In
the rest of the book we discuss the details of how the problem comes
about, ecologically, economically and politically. Hopefully, at the end,
what must be done to protect rain forests will be clear. Hopefully, the
reader will come to agree with us that purchasing tracts of land and
putting them under armed guard is folly. Hopefully, it will become
apparent that stopping individual logging companies and avaricious
agroexporters can be only a small part of the solution, and that basic
questions of land and food security are the most central component of
any potentially effective political strategy. Hopefully, it will be apparent
that such political strategies begin to look more like past political strategies, which helped stop the war in Vietnam and curtailed U.S. intervention in El Salvador and Nicaragua. And hopefully it will be clear that
only by uniting with political forces that have similar fundamental goals
can the future of the world's rain forests be brightened.

---

[1] While Costa Rica's short term debt actually decreased from $575 million in 1980 to $341 million in 1992, probably due to the particular political situation of the 1980's, as described later, its long term debt actually increased from $2,112 million to $3,541 million during that same time period, an increase of 68%. This is not as bad as some other Latin American countries. El Salvador, for example, increased its debt by 208% during that same period, and Mexico managed a 101% increase. Compared to other Latin American countries, Costa Rica is doing better today than it was in 1980, but is still one of the most debt laden of all — worse than Mexico. Only Panama, Argentina, Venezuela, and Nicaragua are worse off.

[2] A hectare is equivalent to 2.47 acres.

[3] The La Selva Research Station, Braulio Carrillo National Park, Tortugero National Park, and the corridor between La Selva and Braulio Carrillo.

[4] Butterfield, 1994; Montagnini, 1994.

[5] Thrupp, 1991.

[6] Most of the history in this section comes from conversations with the late Raphael Echeveria, a long-time local resident and former banana worker and timber cruiser. Information is also derived from Danilo Brenes and Hector Gonzales. The history is effectively the same as presented by Butterfield, 1994.

[7] Lewis, 1992.

[8] In 1989 the president of the Solidarista organization at the Standard Fruit Co. plantations in Rio Frio proudly told one of us (Vandermeer) that "anyone trying to organize a union will be fired. But we know who most of them are and we won't let them get jobs here in the first place." (Paraphrased).

2

◆

The Rain Forest Is Neither Fragile Nor Stable

T he tropics is the area between the Tropic of Cancer and the Tropic of Capricorn. The lowland humid tropics, site of the world's rain forests, account for less than 1/3 of tropical lands — deserts, savannahs, and mountains also occur there. Tropical rain forests are evergreen, or partially evergreen, forests in areas that receive no less than four inches of precipitation in any month for two out of three years, and have a mean annual temperature of more than 24 degrees with no frost.[1] Tropical rainforests are found in three general regions of the world, as shown in figure 2.1. In table 2.1 we present basic data on the extent of rain forests and current estimates of annual loss. Worldwide, tropical rain forests occupy 7.8 million square kilometers of land area,[2] and are being destroyed at a rate of almost half a million square kilometers per year.[3]

Rain forests are enormously complicated creatures. Neither the people who live in them, nor the scientists who study them, understand everything about how rain forests work, which undoubtedly contributes to our mystical feelings about them. Yet comprehension need not elude us completely. Few of us understand how our automobiles work, but we do know certain basics — the motor will start when we turn the ignition, and the accelerator pedal is quite distinct from the brake pedal. In this sense we can also comprehend the way rain forests function. No one knows all the details, and all is not yet understood, but what is known with some certainty can be explained about as easily as explaining how to put the key in the ignition, or when to step on the brake rather than the accelerator.

Most popular accounts of rain forest loss and/or preservation emphasize either a romantic or a utilitarian notion of rain forests, concentrating mostly on the larger political forces driving their destruction. This has led to some confusion, and no little dogma about what can and can not, or should and should not be done to save or restore rain forests.

Rain Forest Regions of the World

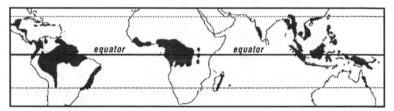

Figure 2.1. The distribution of rain forests in the world.

Table 2.1. Extent of today's rain forests and rate of deforestation.

Region	Current extent	Rate of annual deforestation
America	4 million square kilometers	0.19 million square kilometers
Asia	2 million square kilometers	0.22 million square kilometers
Africa	1.8 million square kilometers	0.05 million square kilometers

The Rain Forest: Stable Web or Fragile House of Cards?

Since ecology began to fire the popular imagination in the 1960's, two major perceptions about the nature of ecosystems have resurfaced with remarkable regularity. These two ideas are in fundamental opposition to one another, yet this conflict frequently escapes notice. On the one hand, ecosystems are thought to be fragile and easily damaged by the careless hand of man or woman. On the other hand, ecosystems are considered stable organisms, honed by evolution to be harmonious; the "balance of nature" is considered a stable equilibrium. If an ecosystem is stable, it is obviously not likely to be fragile. Yet these conflicting notions surface repeatedly in the popular ecology movement.

Perhaps nowhere are these two ideas applied with such misguided fervor as in the popular literature on tropical rain forests. Curiously, both the fragility and the stability of rain forests are attributed to their characteristic high biodiversity. Two common metaphors may help explain the rationale behind this idea. On the one hand, the immense biodiversity in a rain forest may be thought of as the filaments in a spider web; the more connections the stronger the web. Thus, because the rain forest has so many strands and so many connections, it must be very stable. On the other hand, the immense biodiversity may be represented as a house of cards, each card balanced precariously on edge, all the cards supporting one another to keep the house standing. This house of cards may be balanced because of the large number of cards, but removing just one could very well cause the entire structure to come tumbling down.

So which metaphor makes the most sense in fact? Is the rain forest a highly connected web that gains its strength and stability from its multiplicity of connections? Or is the rain forest a house of cards, precariously balanced, and subject at any moment to total collapse if a single card is removed?

It may surprise anyone who has read a few books about rain forests to learn that both these popular visions are probably way off the mark. In the past ten years the science of ecology has learned a great deal about how rain forests work. Added to what has already been gleaned from fifty years of tropical forestry, and from the accumulation of hundreds of years of knowledge of the indigenous people who live in these ecosystems, we now have a relatively clear understanding of the basic functioning of tropical rain forests. To be sure, an enormous number of particulars remain enigmatic — something to be expected from the sheer quantity of elements involved. But the overall picture is probably better understood than might be surmised from a look at the rain forest picture book on Uncle Ed's coffee table. Many of us do not understand how a carburetor works, yet we can easily grasp the function of the accelerator pedal. Similarly, many details of rain forest function are still poorly understood, yet we can fully grasp the nuances of biodiversity and other general features of rain forest function — and perhaps we can use that knowledge to our benefit.

Six Key Factors in Rain Forest Function

The overall picture of rain forest function can be understood with six key ideas. First, as is commonly mentioned in popular accounts of rain forests, there exists an enormous amount of diversity — different kinds of plants and animals and other living things. This diversity in and of itself generates some problems for individual species. For example, if hundreds of distinct species are to exist in the same place, that means most of them must be rare. If there is just room for 200 trees in a plot of land, and you wish to fit in 100 species, each species, on average, can only be represented by two individuals.

This brings up the second factor: sex. How do such rare individuals find mates? The dominant elements of the forest are plants, and especially trees, and sex in plants happens through the process of pollination.

The third key factor is the problem (from the point of view of plants) of herbivores: the many insects and animals that enjoy eating leaves, shoots, and seeds. How plants deal with herbivores is a major piece of the story of how rain forests work. Herbivores tend to specialize on particular species of plants. For example the mahogany moth eats nothing but mahogany trees. If mahogany trees are interspersed with numerous different species, as is the case when most species are rare, mahogany moths will have trouble finding their dinner. If mahogany trees are all clumped together they make an excellent target for herbivores.

This leads to the fourth factor, the dispersal of offspring. If seeds just drop to the base of a plant, and the young plants grow there, any insect that happens to find the adult will find the offspring too. So plants have evolved various strategies to disperse their seeds.

The fifth item is death. In the context of the forest, a dead tree signifies much more than simply the passing away of one individual in a population. A dead tree means a gap in the canopy of the forest. Sunlight may now enter that gap and, for a time, the understory of the forest is quite different than it was before the death of the tree. Thus, a major element of rain forest function is the formation of light gaps by falling trees. What happens in these gaps is of utmost importance to the structure of the forest.

The sixth key factor in rain forest function is the soil. The matrix from which the trees must draw their sustenance is very special for tropical rain forest trees. The same physical forces that make the rain forest itself so incredibly lush also make the soils on which they live rather infertile — an apparent contradiction that we explain below.

So these six factors — high biodiversity, pollination, herbivory, seed dispersal, light gap dynamics, and soils — are the simple elements of how tropical rain forests function. We note in passing that each of these six key elements will play a part in any and all proposed management schemes of tropical rain forest areas. They will be useful in understanding the enigma of poor farming potential in the most lush ecosystem on earth, the headaches of reforestation after commercial logging, and how to make good on the promise of food from the forest.

Biodiversity

We begin by examining the most obvious factor. In a forested area in Northern Michigan we found eight species of trees in a sample of one hectare (approximately 2.47 acres). In the same size area in the rain forest of Nicaragua we have thus far encountered about 200 species of trees.[4] Entomologists netting insects in Kansas found 90 species of insects, whereas applying their nets in the same fashion in a rain forest in Costa Rica they found 545 species.[5] In a Peruvian rain forest entomologist E. O. Wilson identified 43 species of ants in a single tree, almost as many as those reported for all the British Isles.[6] In the Americas, bird species increase approximately five-fold from mid latitudes to the tropics.[7] While there are exceptions to this rule (for example, the number of small rodents running around in the leaf litter of a tropical rain forest is about the same as, or even lower than, the number running around in a northern forest), this general pattern is repeated for most kinds of organisms one examines — trees, herbs, insects, birds, etc.[8]

Ecologists have been debating the significance of the high degree of rain forest biodiversity for a long time, and so far have not come up with a satisfying explanation. On the one hand, there is the persistent and enigmatic question of what causes this great amount of diversity. Initially it seemed that the high productivity afforded by large quantities of rainfall and hot temperatures would promote much diversity. However, when ecologists examined this idea closely it did not prove to be true. Indeed there are convincing arguments that greater productivity could actually result in much lower diversity. From the point of view of how much material is actually produced per unit time, the most productive ecosystem in the world is a modern corn field. Other ecologists have suggested that the longer duration the tropics have been free of the ice sheets that covered much of the temperate zone during the great ice ages has allowed for the evolution of more different kinds of organisms. But here, too, there are opposing arguments and this "time hypothesis" remains, at best, controversial. We could go on with a variety of other explanations, but the story is the same for each one of them. While Newton was able to explain the laws of gravity, and Darwin explained how biological organisms evolved, no comparable explanation yet exists for why some places (like the tropics) have many species, while other places (like the arctic) have so few.

Though the origin of this diversity remains an enigma, the answer to the related question of how all these species are maintained in the same system is slowly being worked out. As mentioned above, the assumption that the more things in a system, the more stable the system has been accepted in the popular literature on rain forests for years. But a careful analysis of how ecosystems work has led us to the surprising conclusion that the fundamental theories of ecology actually predict the opposite. In fact, the more diverse the system, the more likely it is to be fragile. Our house-of-cards metaphor seems more likely to be true than the spider web.

Realizing that a high-biodiversity system, like a rain forest, is more

likely to be fragile than stable, ecologists in the 1980's began trying to figure out how such diverse systems could be maintained. That is, rather
than assuming that such complex systems were stable as a spider web
and asking where they came from, we began admitting that maybe rain
forests weren't so stable and asking how they could persist in the real
world if they were house-of-cards fragile.

This is where ecology stands today regarding the issue of rain forest
stability or fragility. The answers are certainly not yet well established,
but the questions are clearer now than they have been before. Two
research agendas appear to be dominating this issue today. First, many
are asking questions about the particular way in which ecosystems are
held together — the details of the balancing cards, so to speak —
whether a tree with many rare herbivores can coexist in the neighbor-
hood of a tree with a few common ones, or whether a beetle that eats a
bug that eats an herb is more likely to persist in the long run than a beetle that eats a beetle that eats the bug that eats the herb.[9] Second, other
ecologists are focusing on the role of disturbance events such as storms
and landslides.[10] Since such disturbance events are as much a part of
nature as the biological organisms that make up the ecosystem, it seems
important to understand what their ultimate effect might be. How the
diversity arose in the first place is something of an enigma.[11]

Pollination

Since the evolution of flowering plants over a hundred million years
ago, terrestrial ecosystems have been dominated by a fundamental
mutualism — animals help plants have sex and the plants reward the
animals with food. The hummingbird drinks nectar from the flower of
the wild banana plant and takes the pollen from one banana plant to
another. The humming bird gets fed and the banana plant has sex. Both
species benefit from this interaction, the definition of mutualism. While
the vast majority of plants in the world are pollinated by insects, the rain
forest also contains special pollinating systems involving birds and bats.

In a rain forest, the problem of pollination is basically one of being
rare and wanting to have sex (i.e. wanting to reproduce). The problem is

solved in a variety of ways, but most commonly in one of four specific forms. First, particular species may be rare over a large range, but common in a local area. This leaves certain technical genetic problems unresolved and creates problems with elevated herbivory, but solves the simple problem of reproduction. Pollen can reach ova quite easily if a rare species always exists in a clump of its own.

Second, many rare species have evolved a process called *selfing*. Using a variety of mechanisms, many tree species are effectively hermaphrodites. This means that a single individual has the reproductive organs of both sexes. While this seems an obvious solution to the problem of reproduction, it creates enormous alternative complications associated with inbreeding. We all know what happens when cousins marry, and the genetic problems faced by the royal families who inbred are legion. Such problems are magnified enormously when an individual mates with itself. How plants deal with these new genetic problems is well beyond the scope of this introduction.

Third, many pollinating systems for rare plant species include synchronous mass flowering and a long-distance pollinator. For example, some tree species typically display all their flowers in unison, once per year, and all the individuals in the population of that species do so simultaneously. These trees use, as pollinators, bees capable of flying very long distances between meals. Thus, over an extended area, while the trees themselves occur only occasionally amongst a mass of other species, when it is time for sex, all the individuals in the extended population put out all their flowers at the same time. The bees are able to fly from one tree to the other, picking them out amidst the mass of greenery of all the other species because of the abundance of colorful flowers in their crowns.

The synchronous mass-flowering strategy has certain prerequisites. The pollinating bees must be supplied with a food source for the entire year. A particular species of tree will likely only flower for a week or two during that year. In order to have at least one species of tree flowering at all times during the year (an absolute necessity for the bee), there must be a minimum number of tree species in the forest. So, for example, if each of the tree species flowers for two weeks during the year, an absolute minimum of 26 tree species will be needed to maintain this particular form of pollination (assuming any overlap in flowering time is minimal). This requirement has dire consequences for any strategy that attempts to preserve small patches of forest. If, by accident, the patch is missing or loses just one of the necessary 26 species of trees, the whole system could collapse since the bees would die during that two-week period without food, and without their specialized pollinators, the trees would be unable to reproduce.

The fourth solution to the problem of reproduction for rare species is the habit of *trap-lining*. Some birds and insects are capable of creating a mental map in order to remember the locations of particular food sources. An individual plant that is currently producing flowers may be very far from others of its species, but a humming bird, for example, may know exactly where the next individual of that species is located. The humming bird thus has a number of individual plants in mind when she sets off in the morning to find food. She tanks up on nectar at the first site and then flies off to the next plant, remembering from day to day where these particular individual plants are located. She has, metaphorically, a trap line (trappers frequently refer to the series of traps they put out each night as a trap line) formed by the flowering plants she visits each day. The individual plants, in turn, produce only a few flowers each day, attracting the pollinator, but not monopolizing its attention, so that it may fly off to the next individual in the trap line.

These then are the four principle ways in which the problem of sex under conditions of rarity has been solved: local clumping, selfing, synchronous mass flowering, and trap-lining. Of course the details of pollination systems are far more complicated than we indicate here. But the basic patterns are really quite simple.

Herbivory

Herbivory as a way of life has existed for hundreds of millions of years. The spectacular elephants and lithe gazelles of the African savannas, or the cattle herds that graze America's western plains are well known examples of herbivores. While insects are the most important pollinators in tropical rain forests, they are not universally beneficial to plants. As herbivores, insects are probably the plant's biggest threat to life.

Of the many important consequences of herbivory, one that has significance for humans is the evolutionary response of both plant and herbivore. Plants have evolved an overwhelming array of defenses against the many herbivores that potentially attack them. Three major types of defense, have evolved: structural, chemical, and mutualistic.

Structural defenses include all of the spines and hairs that one knows so well from spiny vines, stinging nettles, and cacti. Such structures certainly need no further explanation. But at a more microscopic level, small hairs, what we might normally regard as fuzz, on the surface of leaves, extra thick cuticle on the stem, and a host of other microscopic equivalents of spines have evolved as a protection against insect herbivores. What seems like fuzz to us is an array of stilettos to a tiny insect.

Only recently have chemical defenses been fully appreciated. While we have known for years that many plants are poisonous, we did not realize that most of those poisons had evolved mainly as a defense against herbivores. Indeed, we now appreciate that most plants, especially those living in tropical rain forests, produce chemicals to protect against the ravages of herbivores — natural insecticides, so to speak.

There is, however, a secondary consequence of this chemical defense. Herbivores themselves evolve. And, just as one can often find a chemical antidote for a poison, herbivores have frequently been able to evolve antidotes for the poisons plants produce to protect themselves. This process is so common that ecologists have dubbed it the "co-evolutionary arms race." Every time an herbivore evolves a detoxifying mechanism to deal with a plant poison, the plant is subjected to pressure to evolve a new poison, which puts pressure on the herbivore to develop a new detoxifying mechanism. This seems to be a continuously operative process in nature, and has significant consequences for those seeking to engage in agriculture in rain forest environments.

The third form of herbivore defense is mutualism. In a wide variety of cases, plants have evolved the capacity to use other insects to protect themselves from herbivores. The most celebrated case is that of the ant and the acacia plant. Acacia plants produce hollow spines in which stinging ants live. Further, the acacias provide the ants with a self-contained energy and protein source (the acacia produces glands that produce food exclusively for the use of the ant). In return, the ants act as a kind of armed guard for the acacias. They will viciously swarm over and kill any herbivore that makes the mistake of trying to eat the leaves of their host. This phenomenon is quite common and represents the plant taking advantage of the overall structure of the ecosystem. Every herbivore has natural enemies, parasites and predators that eat it. Structures may evolve to attract and coopt these natural enemies, thus creating a mutualistic relationship between the plant and the predators of its natural enemy, a natural embodiment of the adage, "My enemy's enemy is my friend."

Seed Dispersal

While the problems related to seed dispersal exist all over the world, they are especially important in a tropical forest. Seeds are usually the most energy-rich part of a plant and are thus a prime target for herbivores. The vast majority of energy that humans use is obtained from such seeds as wheat, corn, and rice. If seeds are simply dropped in a bunch at the base of the parent tree, they make an excellent target for herbivorous insects and mammals that eat them. Consequently, most plant species have evolved mechanisms to disperse their seeds away from the fixed position of the parent plant. In tropical rain forests these mechanisms frequently involve animals, especially mammals and birds.

There is something of a contradiction involved in seed dispersal. While it is generally a good, even necessary, process, seed dispersers also have the potential of being seed predators. Squirrels eating acorns, for example, may very well eat all the acorns produced by a tree, leaving none to produce the future generation of oak trees. To solve this problem, plants have evolved two general methods: separation of seed from the disperser's food reward, and satiation. In the vast majority of cases, plants have evolved attractant structures to entice dispersers to scatter their seeds. Almost all of the fruits we eat are examples of this strategy. A bird that eats a berry digests the pulp of the fruit and passes the seeds in its stool. When a horse eats an apple, the lure of the apple is not the tiny brown seeds within, but the juicy pulp — the horse passes the seeds unharmed. The reward for the animal disperser has been separated from the seed itself.

In a significant number of cases the seed is the attractive structure, for example oaks and their acorns. Since the disperser, in this case the squirrel, is by definition also a seed predator, a contradiction emerges, how to provide the attraction, yet get the seeds dispersed. This is actually a rather complicated issue and its resolution is only partly understood. It involves a concept called *satiation.* Oak trees, tend to have so-called mast years, in which an unusually large number of acorns are produced by all the trees in a population. After three or four years of relatively low production of acorns, in a mast year all the trees in a river basin produce an excessive number of acorns. The consequences are not surprising. After a short supply of acorns for three or four years, the squirrel population has thinned out, and all the squirrels in the river basin are simply not capable of eating the large number of acorns produced. They are satiated. In a mast year the animals are satiated as seed predators and are thus able to act as seed dispersers. In the example at hand, the squirrels continue to gather and bury acorns, but have little motivation to go back, dig up, and eat them. Thus they leave a significant number of acorns to germinate and produce new oak trees. This process of seed-predator satiation is sometimes very important in rain forests, and is almost universal among the dominant trees of the Southeast Asian rain forests.

Gap Dynamics and Other Forms of Disturbance

Old forests typically have old trees, and trees do not live forever. A sudden gust of wind will bring an old tree down, creating a hole in the forest canopy, and bathing the understory with light. What follows is that sun-loving plant species, known as *pioneers*, enter the gap first, followed by secondary species, and eventually a *climax* tree grows up to take the place of the tree which fell. Climax refers to the ultimate stage in this successional sequence. Specific kinds of species are associated with each stage in the sequence, the climax species being the ones that are not further replaced until they die. The process of replacing a fallen tree with a new adult tree takes anywhere from thirty to hundreds of years, depending on the type of forest. The continual process of a tree growing, becoming large, falling to create a gap, and eventually being replaced by another, is thought to be a major force influencing how rain forests are structured, and therefore, how they function.[12]

Recent studies suggest that over the last few decades, the turnover rate of these gaps has increased in tropical rain forests.[13] In other words, they form more frequently than in the past. Although the reasons for this increase are not known, it has been suggested that the consequences will be a change in forest structure, and a possible change in function, with an increased predominance of light-demanding plants and the eventual extinction of slow-growing, shade-tolerant species. Plants absorb carbon dioxide and relese oxigen. The carbon in the carbon dioxide is incorporated into the tissue of the plants. Trees with harder wood tend to *secuester* more carbon than trees with softer wood. Light-demanding and fast-growing trees generally have softer wood than the shade-tolerant slower-growing ones. Thus, if the structure of the forest changes from one dominated by slow-growing species to fast-growing ones, the forest will change the rate at which it absorbs carbon dioxide from the air. Since carbon dioxide is one of the main greenhouse gases, this change in forest structure may contribute to the increase in global warming.

In the past few years a great deal of attention has been given to changes in rain forests following destructive storms. A large storm may create an exceedingly large light gap in the forest, and the question arises whether post-storm succession is simply a very big light gap, or whether something qualitatively different happens because of the immense size of the damage done by the storm. This is an important question, and while the answer is still elusive, many studies are currently underway. A storm's damage to the forest is similar in many respects (though not all) to damage done by a logging operation. Understanding the natural processes of how the forest responds to a storm gap or a light gap might very well provide insight for designing more ecologically rational methods of forestry than those in current use.

A final point is worth making here. In the past it has been presumed by romantics that tropical rain forests are ancient places, super stable cathedrals of towering trunks whose age defied human imagination, hardly touched by the hand of *Homo sapiens*, nor ravaged by the vicissitudes of nature. This vision is simply not true. Tropical forests are frequently subjected to tropical storms that periodically knock down almost all the trees, to landslides that remove large sections of vegetation, to naturally-occuring fires, and most importantly in the past several thousand years (at least), to the hunting and agricultural pressure of *Homo sapiens*. Finding a truly "untouched" forest is hardly possible.

This is an important issue. Tropical rain forests can withstand a great amount of physical damage over the long term. They may appear fragil, but we now know they also have a high capacity to restore themselves — perhaps *not resistent* to damaging events, but they are quite *resilient* if given the chance. They inevitably grow back after large storms, after landslides, and after peasant agriculture. The only damage, which may be permanent, results from modern degradations—urbanization and chemically intensive agriculture. A peasant who cuts down a forest to plant corn for a couple of years and then abandons the site, does little long-term damage (the forest grows back). But a banana company that physically alters the soil structure, chemically changes the content of the soil, and saturates the ecosystem with pesticides, has a far greater effect. And a piece of land covered with cement will not recuperate as a tropical forest except over a very long period of time. We return to this theme in a later chapter.

Soils

Plants get most of their nutrients through their root systems. This means that the soil, where the roots lie, is one important determinant of how well a plant does. In understanding the basics of soils it is appropriate to focus on two related but distinct questions. How does the plant get nutrients out of the soil? And, how are nutrients stored in the soil?

How plants get nutrients out of the soil is well understood, largely because of the importance of this topic to the agricultural sciences. Many nutrients occur in the soil as small, charged particles — ions. They are like lint particles with a static-electric charge. The most important nutrients that occur this way are potassium, magnesium, calcium, and one of two forms of nitrogen. All of these have a positive charge, and they are all attached to small clay particles, much like a small magnet attaches to a metal surface. As part of their normal life's activities, plant roots give off positively-charged ions (just as we must excrete urine, plants must excrete certain products of their metabolism). Thus, there is a continual gradient of acidity as you move away from the surface of the root (recall that acidity is simply the concentration of positive charges). Because the plant roots give off positive ions, the zone immediately next to the root surface is more acid than the rest of the soil. When the clay particles with the positively-charged nutrients attached come near to the plant roots they encounter this more acid environment. Because the nutrients are now in a changed environment, they change their chemical form and suddenly become available to be absorbed through the root. It is sort of like the magnets attached to the metal surface are displaced by other magnets, metaphorically the small positive-charged ions that are excreted from the roots. And once they are free of the clay particle (once the magnets disengage from the metal surface), they are available to be absorbed into the root of the plant. It is important to note that it is the gradient of acidity that is important. As explained below, the positively-charged ions are held fast in the soil at an equilibrium level associated with the acidity. It is the sudden change to a more acid state that makes them available to the plant root. The diagram in figure 2.2 explains this process.

The way soils store nutrients is also well understood. Two parts of the soil structure are important: the clays, and the organic matter. Clay's are exceedingly large chemicals (tiny by absolute standards, but huge as chemical molecules go) whose surface is covered with negative electro-static charges. Because of their negative charge they attract the positively-charged ions in the soil. The greater the abundance of this clay component of the soil, the more positive ions will be attached. Here lies the key point. If the positive ions are not attached to anything, they tend to wash out of the system. Thus, it is extremely important to have clay in the soil to keep nutrients from leaching out of the soil.

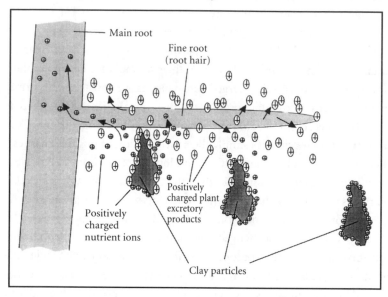

Figure 2.2. Diagram of how nutrients are absorbed by plants. A clay particle (or equivalently a piece of humus) carries with it small charged nutrient ions (small circles with pluses inside). These are the nutrients the plant needs for growth and development. The plant root, especially the very fine tips of the roots, known as the root hairs, excrete other positively charged ions during the course of their normal life's activities. These are the larger ovals with pluses inside of them. As the clay particles come close to the plant root, some of the big positive ions replace the small positive ions on the surface of the clay, thus making those small ions available to be taken up by the plant through the root hair.

Humus, the decomposed organic material of dead plants and animals, functions in a similar way. Very small pieces of humus have negative elec-trostatic charges on their surface and, like the clay particles, they hold the positively-charged ions on their surface. Thus, in terms of plant nutrition and soil fertility, humus acts exactly like the clays, attracting and holding onto positive nutrient ions in the soil, and then releasing them when it comes into contact with the plant roots. But organic matter has an additional use as well. As it decays it releases nutrient ions into the soil. It thus acts as a slow releaser of fertilizer.

Putting together the two concepts of nutrient storage and absorption, it is easy to see the fundamental mechanisms of soil fertility. To the extent that clay particles and humus in the soil have negative charges on their surface, they will tend to take up all the positive charges in the soil, including positively-charged nutrient ions. But when one of these clay or humus particles is washed close to the surface of a plant root, a more acid environment is encountered and some of their positive nutrient ions are replaced with positive ions the plant has excreted. The positive nutrient ion is thus freed from the clay particle, and free to be absorbed through the root of the plant (see figure 2.2). Note how important the acidity gradient is to the whole process. If, when a clay particle moves close to the plant root, it is not induced to give up some of the nutrient ions sticking to its surface, those ions will not be available to the plant. And if there is no gradient (for example, if all of the soil is very acid so that the zone next to the surface of the root is no different), none of those nutrient ions will be available.

That, then, is the general picture of soil fertility and plant nutrition. First, clay particles and humus hang on to the nutrients, preventing them from washing out of the system. Then, when the clay and/or humus particles come within the vicinity of a root surface, they give up the nutrients they are carrying so the plant can absorb them.

Tropical rain forest soils have several characteristics, which make them problematic.[14] First, they are highly acid. This means that the critical acidity gradient between the general soil and the root surface is less dramatic than in other soils. Thus, any crops that farmers try to grow tend to have problems getting the proper nutrients from the soil. Second, rain forest soils typically have the type of clay particles that bear few negative charges on their surfaces and are thus not capable of storing very many nutrient ions. Whatever nutrients may be present, even those from added chemical fertilizer, tend to rapidly leach out of the soil. Third, because of continually high temperature and abundant moisture, the process of decomposition occurs rapidly and the organic matter thus disappears rapidly when the natural forest is taken away. This means that both the source of nutrients and the storage capacity of the soil, consequences of soil organic matter, are very poor when farmers try to grow crops.[15]

The Diversity of Tropical Rain Forests

A fact often not fully appreciated in the popular imagination or literature, is that tropical rain forests not only contain a great deal of biodiversity, but there is an enormous diversity amongst kinds of tropical rain forests as well. In a single region one may find distinct combinations of plants and animals on ridgetops as compared to valley bottoms, or on well-drained soils versus humid soils. Indeed it is the bane of the lumber operator in the American tropics that some patches of forest contain particular useful species of trees, while other patches contain mainly species that have little market value.

Most important at this level of diversity is the distinction between major groups of forests. In figure 1.2 we illustrated the location of the world's major tropical rain forests—the Americas, Africa, and Southeast Asia. The forests in these areas do not operate by universal rules, and indeed, sometimes differences between them are extremely important for the practical problems of forestry and agriculture.

For example, many of the forests of Southeast Asia are dominated by a single family of trees, the *Dipterocarpaceae*. In these forests this family may comprise up to 80% of the canopy trees and 40% of those in the understory, whereas in Africa there is only a handfull of tree species that belong to this family, and in South America, perhaps just one.[16] The *Dipterocarpaceae* family contains tree species that typically have straight trunks, grow to a very large size, and can usually be converted into valuable timber. Because of their generally large size and uniformly straight trunks, they are a timber company's dream. In a single hectare of Southeast-Asian rain forest, one might encounter 200 species of trees,

180 of which will be in the family Dipterocarpaceae. By comparison a similar plot in America or Africa might also contain 200 species, but they may be members of 150 families, and the most common family might contain only 10 species. Thus, almost any area of a dipterocarp forest can be utilized for cutting, as compared with the American or African tropics where an area must be carefully scouted ahead of time to locate patches of valuable timber.

Furthermore, Southeast-Asian forests have a very different physical aspect as compared with American rain forests. The understory vegetation in a Southeast-Asian forest is composed mainly of small trees; the seedlings and saplings which will eventually grow up to become the large trees of the forest. In America, this vegetational stratum is dominated by plants that live only in the understory. This understory is where a wide variety of ornamental palms and other species come from. They never grow very large and live comfortably in a low-light environment such as is found in a rain forest understory, doctor's office, or suburban shopping mall.

Such differences suggest that the details of ecology are also quite discrete from place to place, and the more we learn of rain forests the clearer it becomes that this is true. It is becoming increasingly difficult to write a chapter such as this, which makes generalizations that apply to all tropical rain forests. Professional ecologists rarely refer any longer to "the tropical rain forest," but rather specify whether they are talking about an American, African, or Asian forest.

The Biological Side Summarized

This chapter is a brief outline of some of the biological/ecological concepts necessary for understanding the rest of our book. It is useful to understand how a rain forest works, and certainly it will be helpful to comprehend both what happens, and why, when a rain forest is converted to agriculture. As will become clear in later chapters, knowledge of these issues will also be constructive as we try to come to grips with the difficulties of developing sustainable logging methods for rain forests. Finally, the political, economic, and social forces that are really behind the destruction of the world's rain forests are made clearer with the help of this background.

By now, hopefully, we have demonstrated that neither of our metaphors is perfectly accurate. Rain forests are not terribly fragile, nor are they especially stable. Rain forests are, however, highly complex and require a great deal of detailed plumbing and wiring and housekeeping — pollination systems, seed dispersal, defenses against herbivores, and proper soil conditions — in order to function properly. Rain forests may even require periodic natural disasters, such as storms and landslides, and they are attuned to the recurring formation of light gaps. And while the plants of the rain forest live in an Eden of heat and water, they suffer from a virtually empty pantry of the nutrients they need to survive.

---

[1] Majer, 1980.
[2] Whitmore, 1991.
[3] World Resources Institute, 1994.
[4] Vandermeer et al., 1995.
[5] Janzen and Schoener, 1968.
[6] Höldobler and Wilson, 1990.
[7] Terborgh, 1992.
[8] Huston, 1995.
[9] Werner, 1992; Schoener, 1993.
[10] Yih et al., 1991; Mooney and Gordon, 1983
[11] Ricklef, R. E., and D. Schluter, 1993.
[12] Denslow, 1987.
[13] Phillips and Gentry, 1994.
[14] Richter and Babbar, 1991.
[15] However, as discussed in the next chapter, if there is a general characteristic that applies to all tropical soils, it is their heterogeneity. Having most of their experience in temperate regions of the world, many soil scientists viewed tropical soils as uniformly acid and infertile. More recent studies have emphasized that this view is incorrect (Sanchez, 1976; Richter and Babbar, 1991), that soils vary from highly fertile to highly infertile in the tropics. On the other hand, for most regions of the tropics, agricultural conversion has already discovered the good soils, as discussed in Chapter 3, and the remaining rain forests are generally on very poor acid soils.
[16] The species of this family found in South America is a recently described species, which is placed in its own subfamily, but its relationship with the Asian and African dipterocarps is somewhat enigmatic. (Mabberley, 1992).

3

◆

Farming on Rain Forest Soils

The popular perspective that the logger's chain saw causes rain forest destruction is incomplete. There are complex social and ecological forces involved in agriculture that are of far greater significance. Furthermore, socio-political factors are often of overwhelming importance. It is crucial to understand both the technical/ecological issues, *and* the socio-political ones. In this chapter we present the relevant technical issues in a highly abbreviated form. In the two chapters that follow we cover the socio-political factors.

At the outset, we must acknowledge the temptation to assume that in rain forest areas the potential for agriculture is great. Since there is neither winter nor lack of water, two of the main limiting factors for agriculture in other areas of the world, it is easy to conclude that production might very well be cornucopian. The tremendously lush vegetation of a tropical rain forest only heightens this impression, and indeed, this perception may ultimately be valid. Being able to produce for twelve months of the year without worrying about irrigation is definitely a positive aspect to farming in such regions. But, so far at least, the woes are almost insurmountable, as most farmers forced to cultivate in rain forest areas can attest.

The first problem is with the soils. As described in Chapter 2, rain forest soils are usually acidic; made up of clay that cannot exchange nutrients well; and very low in organic matter.[1] Even if nutrients are added to the soil they will be utilized relatively inefficiently because of the acidity, and then they will be washed out of the system because of its low storage capacity.

This problem is actually exacerbated by the forest itself. Because tropical rain forest plants have faced these poor soil conditions for millions of years, they have evolved mechanisms for storing the system's nutrients in their vegetative matter (leaves, stems, roots, etc.) — if they did not, much of the nutrient material would simply wash out of the system and no longer be available to them. This means that a vast majority of the nutrients in the ecosystem are stored in plant material rather than in the soil. This is exactly the reverse of the corn-belt soils of the United States. Indeed the cycling of nutrients in a tropical rain forest is considered highly efficient, with surface roots absorbing nutrients as soon as they are released from decaying matter.

The consequence is that when a forest is cut down and burned, the nutrients in the vegetation are immediately made available to any crops that have been planted. The crops grow vigorously at first, but any nutrients unused during the first growing season will tend to leach out of the system. The "poverty" of the soil only becomes evident during the second growing season. This pattern is especially invidious when migrants from areas with relatively stable soils arrive in a rain forest area. The first year they may produce a bumper crop, which gives them a false sense of security. Then, if the second year is not a complete failure, almost certainly the third or fourth is, and the farmer is forced to move on to cut down another piece of forest.

A second problem is pests in the form of insects, diseases, and weeds. The magnitude of the pest problem is often not fully anticipated by farmers or planners, and it is only after problems arise that the surprised agronomists become concerned. This is unfortunate, since one of the few things we can predict with confidence is that when rain forest is converted to agriculture, many pests arrive. The herbivores that used to eat the plants of the rain forest are not eliminated when the forest is cut. They are representatives of the massive biodiversity of tropical rain forests, and their potential number is enormous. Herbivores can devastate farmers' fields, and are able to destroy an entire crop in days.

A third problem is that because of the uniformly moist and warm environment, organisms that cause crop diseases find rain forest habitats quite hospitable. Consequently, the potential for losing crops to disease is far greater than in more temperate climates. Finally, just as the hot, wet environment is agreeable for crops, it is also agreeable for competitive plants. Since no two plants can occupy the same space, frequently the crop falls victim to the more aggressive vines and grasses that colonize open areas in tropical rain forest zones. Weeds are thus an especially difficult problem.

Rain Forest Landscapes, the Invisible Mosaic

Like other ecosystems, tropical rain forests exist in diverse settings. These settings frequently set the stage for the details of agricultural incursion into an area. In particular, the nature of the soils is governed by the same forces that govern the formation of the landscape. In any rain forest area we can expect to find a variety of basic soil types. There are five basic soil types of relevance to agriculture in rain forest areas: acid soils, alluvial soils, volcanic soils, hillside soils, and swamp soils.[2]

Acid soils are the classic rain forest soil. When the main bulk of the soil is acid, the acidity gradient between the general soil and the area immediately adjacent to the plant root virtually disappears. Since this gradient is essential for plant nutrition,[3] acid soils are very poor when it comes to feeding plants. Because of this acidity, even adding nutrients to such soils does not make much difference. Furthermore, acid soil usually contains a very small amount of organic matter, and the sort of clay minerals with low negative charges. Thus, the nutrient ions that do get into the soil are rapidly leached out. In sum, acid soils contain few nutrients to start with, and those contained are frequently unavailable to the crops anyway.

The origin of these acid soils is not mysterious. When a soil first comes into being, the clay particles[4] it contains are large and incorporate a great deal of aluminum and iron within their chemical structure. The large surface of each clay molecule provides many sites for negative electrical charges. These negative electrical charges attract positive ions dissolved in water in the soil, thus making the soil less acid[5] and keeping the positive ions around where they can later be used as nutrients by the plants. But weather effects, especially high temperature and moisture, cause the clay molecules to decay. As they decay they become smaller. As they become smaller there are fewer sites for negative charges, and some of the aluminum and iron that was contained within the clay molecule is released to the soil in the form of ions.[6] Since these are positive ions, they contribute to increasing the acidity of the soil. Because the smaller clay molecules can no longer neutralize as many positive ions, this also contributes to the soil becoming more acid. This process, known as *weathering*, is what makes acid soils.

Alluvial soils are deposited on the floodplains of rivers. As a result of the abundant living creatures in rivers, alluvial soils tend to carry a lot of organic matter, which gets deposited on land during floods. The mud left after a river floods may look ugly and do immense damage to a living room, but it contains large quantities of biological materials. Because of this organic matter, alluvial soils can store nutrients quite well.[7] The Mississippi and Rhine Valleys are excellent examples of areas with rich alluvial soils, rightfully famous for their agricultural productivity. Wherever there are rain forests, one can almost always find alluvial soils along the borders of the rivers that flow through them. One can frequently draw an approximate map of where the alluvial soils are, simply by marking all the agricultural fields, as farmers can be quite good at finding the best soils in an area.

The third major soil type is derived from volcanic ash. Volcanic soils do not always occur in rain forest areas. Indeed, they are an exception rather than the rule, but when they do occur they are important agriculturally. They are usually extremely productive, sometimes rivaling the rich corn-belt soils of North America. As they age they too become acid through the weathering process. But such weathering requires hundreds or thousands of years, and while they are young, volcanic soils do not have the undesirable qualities of acid soils. Young volcanic soils are particularly good for agriculture because the clays they contain are well-known for their ability to capture and retain nutrients. In some of the rain forest areas of Java and Hawaii, for example, volcanic soils are very fertile. The beautiful Javanese rice terraces would likely still be rain forests, had not the rich volcanic soils led farmers hundreds of years ago to cut the forest and replace it with rice fields. For obvious reasons, rain forests growing on these rich volcanic soils throughout the tropics were the first to be cleared for agriculture.[8]

Hillside soils have one important characteristic. They erode very rapidly. The natural vegetation that covers them is effectively the only protection they have against severe erosion. When converted to agriculture, hillside soils are rapidly eroded; the top soil washes away and they soon become unproductive. Unfortunately, because of economic and socio-political pressures, many peasant farmers are forced to farm on these soils with inevitably poor results.

Finally, many tropical areas encompass natural wetlands, whose soils — in the U.S. they are called swamp or muck soils — are extremely rich in organic matter. However, except for rice and a few other crops that can grow under innundated conditions, it is not possible to do agriculture in a swamp. But with large-scale projects, swamps can be drained and the resulting soils are extremely rich. This, however, is not an activity normally undertaken by small farmers. Rather it is more likely to be associated with larger capitalized operations like those of the big fruit companies.

In light of the above, it is oversimplified at best to state that all tropical soils are poor. There is a range of fertility associated with tropical soil types, which depends upon local conditions. From the extremely rich volcanic soils, to the moderately rich alluvial soils, to the rich but hard-to-manage swamp soils, to the extremely poor acid soils, to the hillside soils that erode so easily, there is inevitably a mosaic of soil types, and consequently a mosaic of agricultural potentials that exists in rain

forest areas. Note the distribution of soil types in an imaginary landscape
that could arise as we depict them in figure 3.1.

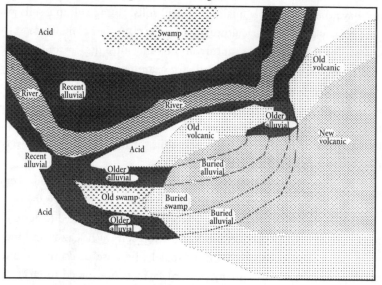

Figure 3.1. Hypothetical illustration of typical distribution of soils under a tropical rain forest. The river running through the forest deposits the recent alluvium
every time it floods. The river used to have a different channel and used to
deposit alluvium in that channel before natural erosion changed its course. This
is what made the old alluvial deposit. Part of that older alluvial deposit was
buried by a recent volcanic eruption, and thus is called *buried alluvial.* An earlier
volcanic eruption deposited ash in a larger area, and before the river changed its
course. Thus the older alluvial soil is actually on top of the soil that was produced
from the earlier volcanic eruption. The soils that were never covered with either
volcanic ash or alluvial deposits are the oldest soils and are very acid. A small
swamp used to exist just north of the river, creating a humus-rich swamp soil.
From a farmer's point of view the soils range from the really excellent young volcanic soil to the recent alluvial, to the swamp, to the old volcanic, to the old alluvial, to the acid. At one extreme (new volcanic) the soils are quite good, but at the
other extreme (acid) the soils are among the worst in the world. And the trick is
that you can't necessarily tell where all of these soils are since everything other
than the river is covered with rain forest.

Without knowing the history of such an area (that there is an old river
channel, that part of it is covered with an old volcanic ash deposit, etc.),
one would be hard pressed to know exactly where the good and bad soils
were located. There is an unseen mosaic of soils underneath any rain

forest, some favorable for agricultural production, others less so. In figure 3.2 we present a production plan based on this soil mosaic. However, without a detailed study, something the peasant agriculturalist is not likely to be able to afford, one cannot tell where those good soils lie, and our plan, however rational ecologically, would be impossible to formulate under current political conditions. Thus, most actual agricultural planning in rain forest areas is done in relative ignorance of the underlying soil mosaic. This problem is faced by both the small farmer and the planner who seeks to promote efficient agriculture in such areas.

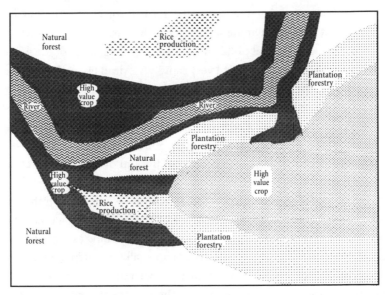

Figure 3.2. A production plan based on the underlying soil mosaic. Each cropping system is based on the land-use capability. Note how such a system would not conform well to the notion of straight-line fences and rows.

Yet even though this unseen mosaic is a reality in all rain forests, it remains true that a large proportion of the soils under current rain forest cover are of the highly acid type. Indeed, in many areas of the world, the volcanic and alluvial soils have already been converted to agriculture and only the acid soils and soils on very steep slopes remain covered with forests.

Slash-and-Burn Agriculture

The most common form of traditional agriculture, and one still practiced many places in the world today, involves the use of fire, and a
reliance on the ecological process of <u>succession</u> (figure 3.3). In principle,
slash-and-burn agriculture represents the easiest solution to two ecological problems inherent in agricultural production: the <u>problem of</u> plant
competition (weeds) and the <u>problem of nutrient</u> cycling. If crop seeds
are scattered in a patch of cleared forest, the plants already there will have
a clear advantage over the newly seeded ones. By burning the cleared
patch before scattering the seeds, the farmer forces the undesirable
species to at least begin at the same level as the crops.

The second ecological problem is <u>nutrient cycling</u>. Most of the nutrients in a rain forest are recycled very rapidly. As dead leaves decompose,
they release their nutrients into the soil where they are then picked up by
the roots of living plants and reused. We have reason to believe that most
systems undisturbed by *Homo sapiens* are in relative balance regarding
nutrient cycling — the amount of nitrogen, for example, released into
the soil by decomposing leaves equals the amount of nitrogen picked up
by the trees as they put on their new leaves.

But consider what happens when agriculture is imposed on the system. Harvesting means removing some of the biological material from
the system. It is not then available for recycling. Burning exacerbates this
problem tremendously in that nutrients normally stored in plant material — thus held securely in the system — are <u>released en masse</u> by the
<u>burning</u> process. Crops utilize some of these nutrients, but the rest may
be taken out of the system through water runoff or leaching. After a
while, since the nutrient-cycling process is interrupted and there is a <u>net</u>
<u>export of nutrients</u> out of the system, the land must be abandoned.

Subsequent to abandonment, the process of ecological succession
takes over and various plant and animal species occupy the abandoned
agricultural land, slowly bringing it back into a relatively stable, nutrient-
cycling system. After a <u>fallow period</u>-during which the patch has
returned to forest, the farmer may return, slash and burn the new vegetation, and start the cycle over again. Depending upon the nature of the

soil, and of the successional process that takes over after the land is left idle, this cycle may be as short as ten years, or as long as fifty. In a sense, ecological succession "solves" the problem of the disruption of the nutrient cycling system caused by agriculture.

Figure 3.3. Slash-and-burn agriculture. Top photo is of a recently slashed and burned area in Tabasco, Mexico. Bottom photo is the same area after crops have begun growing.

The slash-and-burn system can be seen as cutting and applying fire to natural vegetation to release the desired plants (the crops) from biological competition, and to provide them with large amounts of nutrients. However, slash-and-burn creates a problem of net nutrient loss from the system, which requires abandonment of a particular site after a short period of time (two or three growing seasons) to allow the process of ecological succession to bring the system close to its original pattern of nutrient cycling.

Stabilizing Agriculture on Rain Forest Soils

Before beginning this section, we wish to remind the reader that the reasons most small farmers move their farms into rain forests have much more to do with politics than with nature, a point to which we return in the next chapter. But even if the socio-political landscape were utopian, the biophysical conditions of rain forests make them difficult to farm in perpetuity. At least that is the case with current forms of peasant agriculture. On the other hand, it is also true that some indigenous people have farmed "within" rain forests for generations without major disruption of the overall forest landscape.[9] Sometimes these indigenous populations are very small and practice slash-and-burn agriculture at the margins of old rain forest, allowing their fallow only ten or twenty years of succession before returning to it. More rare are those groups who actually farm within an intact forest. We know of few such examples.[10] Most instances of indigenous people "farming" the rain forest are really examples of shifting agriculture in which the older fallow looks much like an old growth forest. While the lessons we may gain from such systems are important and potentially useful to non-traditional people, we emphasize that such populations represent a very small fraction of the agriculture currently practiced on the margins of rain forests.

Given the above, it is important to ask what might be done, technically, to stop peasant agriculture from incursions into rain forest areas. That is, even in the bright hopeful future when small farmers have full title to their land, the full security of credit, and the availability of fair markets;[11] still the technology currently available for producing basic grains on rain forest soils is not sustainable — the soils become depleted of their nutrients rapidly; weeds build up to unacceptable levels; and frequently pests and diseases become uncontrollable, forcing the farmer to move onto a different piece of land. Ultimately we must deal with this problem.

The first and most obvious solution is simply to restrict agriculture to the patches of relatively good soils that exist in rain forest areas. To a great extent these were precisely the areas cleared for agriculture in the first place. For example, in some areas of Central America one can almost draw a map of where the patches of good soils are located, simply by mapping the banana plantations. These plantations have been placed, for the most part, on alluvial soils. A quick glance at figure 3.1 suggests that if we had a rational society we would designate the new volcanic and the recent alluvial soils for agricultural purposes, and probably devote the rest of the soils to forestry or simply let them remain as natural forest, as suggested in our action plan of figure 3.2. But in in a time when rational societies are apparently regarded as utopian, we must look for more practical solutions. We are thus left with the problem of farming on acidic rain forest soils.

Based on proven technology there are three general principles that should be followed in planning agriculture in tropical rain forest areas. First, do not attempt basic grain production on highly acidic soils.[12] Second, grow wood (trees) on the poorest soils. Third, incorporate perennial crops (crops that do not die after a single year of production, such as cooking bananas, or fruit trees) in whatever agricultural activities are proposed. These points require little elaboration. Except for the traditional method of slash-and-burn, which requires a large amount of land per person, the prospects for productive agriculture on rain forest acid soils are not great. Future technologies may change this assessment, and someday we may be able to take advantage of the twelve-month growing season and abundant water supply. However, with today's technologies, the prospects are quite dim. The ecosystem that best develops on acid soils is rain forest, with trees as its main structural component. Ultimately we expect it will turn out that growing trees on acid soils is the best use to which they can be put.

Agricultural activities begin to look more attractive in the patchwork of old swamp soil, old alluvial soil, old volcanic soil, or any other idiosyncrasy that produces patches of less acid (and therefore potentially better) soil. It is here that some of the successes of intensive agricultural production occur in rain forest areas. Examining these successes, one is struck by a single common feature: they are almost always based on perennial crops, or have perennial crops as part of the overall system.

For example, Daniel, a farmer we know, has been farming the same patch of land in Nicaragua for twenty-five years. His corn production is based on a two-year cycle, with a fallow year between each production year. The fallow is largely dominated by two particular species of plants, a small banana-like herb known as "wild banana" and a morning glory vine. Daniel argues that these plants are an important part of the overall system since they keep the grasses out. According to Daniel, the grasses "burn the corn," (by which he means they are bad weeds that reduce corn production). If a patch of fallow starts showing signs of grass, he immediately cuts the grass out by hand. When he is ready to plant the corn, he simply cuts down the morning glory vines and the wild banana, and plants the corn amidst the mass of dead leaves and stems. As the corn grows, so do the vines and wild bananas, presumably choking out the grasses which would "heat up the corn" too much.[13] Amidst Daniel's farm is a virtual forest of fruit trees. Coconut, mango, cashew, and a host of other fruit trees dominate the agricultural system. The corn production is accomplished in the "gaps" within a forest of fruit trees, which dominate the overall agroforestry system. When Daniel talks about his farm, he mainly talks about the fruit trees.

Similarly the celebrated Javanese home gardens are famous for the high diversity of fruit and other trees that inevitably form the backbone of the production system. The gardens are small areas (usually less than one hectare) of mixed trees and crops adjacent to the house. They are characterized by a multistoried vegetation structure and a high diversity of plant species, resembling the configuration of a tropical forest. More than 600 species of plants are known to be grown in these home gardens.[14] They normally include staple food, such as taro or cassava, a vari-

trees = key
agriculture
Also perennials

ety of fruit trees and vegetables, spices, medicinal plants, herbaceous plants, and trees used for building materials and firewood, and finally, cash crops such as coffee, cacao, pepper, oilpalm, and even some timber species, such as teak. Many home gardens also have domesticated animals, such as chickens and pigs, as well as fish grown in ponds.[15] Given the variety of food and fiber produced in these home gardens, it is easy to appreciate their importance for the nutrition of Javanese families. Although home gardens are most conspicuous in Java, occupying 15- to 75% of the cultivated land,[16] they can also be regularly found in other tropical areas of Asia, Africa, and Latin America. They are always located on very small parcels of land, intensively managed, with a wide diversity of plants and animals. But the key feature is still the predominance of trees in the system.

Even in one of the most notorious agroecosystems in rain forest areas, the predominant crop is a perennial. The banana plantations of Central and South America are based on a perennial crop. While all the banana stems are cut once per year, they are not killed, and each banana stem resprouts into a new tree. Thus the basic idea of maintaining a structured root system in the soil is an essential feature of even this high-tech system. Other cash crops now infamous in tropical rain forest areas include cacao (chocolate), rubber, oil palm, and a host of tropical fruits, all perennial crops.

The success of agroecosystems based on perennial crops is not yet fully understood, but one simple explanation makes intuitive sense. Since the main problem is nutrients being exported from the system during times when no crops are covering the ground, the placement of permanent plants in the soil provides a network of roots to "catch" the nutrients before they run out of the system. When annual crops are the main feature of the system, the soil is bare for part of the year allowing the abundant rainfall to percolate through the soil. As it percolates it washes away the nutrients. With living roots permanently in the soil, much of the water is absorbed into the plant tissue and thus, does not simply wash through the soil. The net result is that perennial crops tend to reduce the loss of nutrients from the system.

Some Models for the Future

For the immediate future we see little hope for major advances on the technical front. What is easily knowable seems already known, if not obvious — use of perennial crops, trees, and restriction of agricultural activities to the patches of relatively good soils. Since many farmers who are currently forced to farm in rain forest areas are not utilizing these obvious methods, a great potential exists for improvement, with little or no technological development. For the most part it is a question of political will, a topic to which we return in later chapters.

On the other hand, some intriguing theoretical models are available. Gómez-Pompa, for example, based an ideal model of agriculture in lowland tropical Mexico on the presumed production technology of the ancient Maya.[17] The Chinampa system, widely used by the Aztec cultures in highland Mexico, is a complicated integrated system of agriculture and aquaculture, based on raised fields and the physical manipulation of large amounts of organic matter. To the extent we understand how the Maya used what appears to be a similar system in the lowland tropics, it represents a system of tight nutrient cycling in rich swampy regions.

In figure 3.4 we summarize the operating principles of the Chinampa system, at least those that remain functioning in the central valley of Mexico today. Terrestrial platforms are constructed from swamp soils, creating a network of canals. Biological activity in the canals (photosynthesis, herbivory, and decomposition) utilizes the nutrients in the water and incorporates them in biological material in the form of aquatic plants, fish and other aquatic animals, and the rich organic muck at the bottom of the canal. The fish and other aquatic animals can be harvested directly for human consumption. The aquatic plants can be harvested and used as a mulch on the terrestrial platforms, or composted and incorporated as organic matter. The organic muck at the bottom of the canal is used to form seed beds. Agricultural production on the terrestrial platforms thus utilizes the materials from the aquatic ecosystem to improve the quality of the soils.

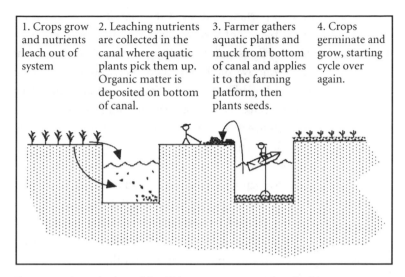

Figure 3.4. Functioning of the Chinampa system, as described in text.

But the key to the system is the nutrient runoff problem. The Chinampa system solves this problem elegantly, not by trying to stop nutrient runoff, but by expanding the system to include a part that literally catches the nutrients as they run off. The aquatic vegetation utilizes the nutrients that escape the terrestrial agricultural system and reincorporates them into the biological material. The farmer then physically returns those nutrients to the agricultural component of the agroecosystem. Note that the "agriculture," the part of the system that actually produces the crop, is only a part of the "agroecosystem." The agroecosystem includes both the agricultural platforms and the canals.

Another example, not currently in use in a lowland rain forest region, is the soil-management practices of the Mayans in the Almolonga Valley of Guatemala. Here farmers utilize the leaf litter of the forests in the surrounding hills as an organic mulch and incorporate it into the soils as they prepare the land for planting. In those hills one can find various combinations of oak, pine, and tropical forest vegetation. Depending on the crop and the perceived condition of the soil, various formulas for combinations of litter are utilized. Production appears to be high and sustainable.[18]

One of the most important features of the Almolonga system is the conception the farmers have of their production unit. The agroecosystem is not simply the fields in which they produce their crops. The Almolonga farmers conceive of their agroecosystem as including the forests on the surrounding hillsides, as much as the valley bottom land where their fields are located. While the trees may not be located exactly in the fields, their presence is recognized as extremely important to the function of the overall system.

Unfortunately these examples are quite theoretical. Although both are functional agroecosystems, they are not located in lowland, wet tropical areas. While the archaeological evidence is relatively clear that the Chinampa system was utilized in one form or another by the Classic Maya, it is no longer, and we really do not know the details of how the Maya actually managed it. Despite valiant efforts thus far, attempts at reconstructing Chinampas in tropical areas have not met with much success,[19] and we are left with the unfortunate conclusion that both the Chinampa and Almolonga systems are simply intriguing models, which someday may form part of the solution to the problems of agriculture in rain forest areas.

---

[1]There is some debate about the question of organic matter in rain forest soils. The rate of decomposition of organic matter is about twice the rate it is in a normal temperate zone situation, and thus it is only natural to expect the standing crop of organic matter to be less in the rain forest. Some authors have questioned this basic assumption (e.g. National Research Council, 1993). On the other hand, all are in agreement that once the forest is cleared for agriculture, whatever organic matter was actually there rapidly disappears from the soil.

[2]For anyone interested, the USDA system would classify the acid soils as either *ultisols* or *oxisols*, the aluvial soils as either *inceptisols* or *entisols*, the volcanic soils as *anidsols*, the hillside soils under a variety of categories, and the swamp soils as *histosols* (USDA, 1975).

[3]This point was discussed in Chapter 2.

[4]Clay particle in this sense is a crystal chemical molecule. The molecule itself becomes exceedingly large since it is built up on the basis of a repeating system of strong chemical bonds.

[5]Remember, acidity is simply the proportion of positive charges in the solution.

6The red color of many tropical soils is due to this excessive amount of iron.

7Recall that humus, decaying organic matter, acts physically like a clay particle, with net negative charges on its surfface, as discussed in Chapter 2.

8Westoby, 1989.

9Shiva, 1993.

10Many examples that have been cited in the past, such as the Bora and Kayapo of the Amazon (Altieri and Hecht, 1990), the Meskito of Nicararagua and Honduras (Neitchman, 1973), and Maya of Mexico (Gomez-Pompa et al., 1987) do not really farm inside of a rain forest. They cut a piece of rain forest down in order to do slash-and-burn agriculture. While it is true, as we argue in Chapter 5, that this form of agriculture is really not very destructive, it nevertheless is agriculture in an area in which the trees of the tropical forest have been cut down.

11These ideas are fully explored in Chapters 4 and 5.

12Recently Huston, 1994, has made this point from a strictly empirical point of view, focusing on the preservation of biodiversity per se.

13This idea, which we suspect is far more common than the literature would suggest, has been explored in coniderable detail by Chacon and Gliessman (1982) in the context of the *buen monte* (good weeds) versus *mal monte* (bad weeds) system of the lowland Maya in Mexico.

14Brownrigg, 1985; Soemarrvato, et al., 1985..

15Personal observations and Michon, 1983.

16Stoler, 1978.

17Redclift, 1987.

18Wilkin, 1987.

19Chapin (1988) has taken to task those who would propose models as tried and true technology. While it is true that the Chinampa system, for example, is an elegant theoretical system, much research remains to be done before it can be regarded as a viable alternative.

4

◆

The Political Economy of Agriculture in Rain Forest Areas

s described in Chapter One, during their normal operations
loggers open up roads, cut many of the larger trees, and do
extensive secondary damage to rain forests. Though there is evidence that tropical rain forests are able to withstand this sort of intervention — they have an amazing recovery capacity — that capability is significantly reduced when the forest is converted to agriculture.[1] The most damaging kind of agriculture for forest regeneration is the modern capital-intensive system. To understand this system fully, it is essential to learn how it came to its present configuration and how it functions. In this chapter we trace the development of agriculture from its early origins to its current modern export base. In the chapter that follows we discuss how agriculture operates today, in both the developed and the underdeveloped world.

We begin with three separate but related topics: 1) the origin and intensification of agriculture; 2) the origin and spread of enclave production, a form of agricultural social organization important in the world's tropical regions; and 3) modernization, especially as it relates to machines, chemicals, and food processing.

The Origin and Intensification of Agriculture

In the previous chapter we touched upon the slash-and-burn system, emphasizing its ecological properties. Undoubtedly the first well-developed form of agriculture, slash and burn probably appeared for the first time in the Mediterranean. While this system is usually practiced in savanna or scrub-like vegetation, it was also practiced in tropical rain forests occupied by human beings. Thus, slash-and-burn agriculture represents the first significant human activity that can be thought of as *deforestation*.

As with all human practices, slash-and-burn agriculture is continually evolving. One pattern of that evolution is a tendency to shorten or lengthen the fallow cycle (i.e. the period when the plot is left idle and secondary succession occurs), depending upon conditions. Thus, a twenty-year cycle might evolve into a sixteen-year cycle and then a ten-year one, or it could evolve the other direction into a twenty-five-, and then a thirty-year cycle. When fallow time is shortened, it is known as *intensification* of agriculture. *Extensive* systems involve large amounts of land in fallow and gradually give way to *intensive* ones in which all the land is simultaneously under cultivation.

Earlier agricultural analysts emphasized a sort of Malthusian approach to this topic, suggesting that it was necessary to put increasing amounts of land into cultivation in order to feed an ever-growing population, thus leading to an inevitable shortening of the fallow cycle. There was thought to be a simple relationship between agriculture and population growth, with population growth the underlying driving force, and increased agricultural production allowing population to reach its more-or-less natural high level.

This view was discarded some time ago. The labor required to make land agriculturally productive, is large, and frequently it is agricultural production itself that promotes a faster rate of population growth, precisely the opposite of what had previously been assumed. If a farmer sees that a certain number of acres must be planted to feed her family, she understands that a certain number of work days are required to make that piece of land productive. If the number of work days is considerable, it is natural to hope for sons and daughters to help with those hours. Population growth is thus promoted. That these sons and daughters will some day require yet more land to feed them, is not likely to come to mind during the long hours of back-breaking work in the field.[2]

On the other hand, it is almost certain that the opposite occurs also. The labor necessary to feed a family may be small due, for example, to a small family, or highly productive soil. The farmer will not need to work many hours. The tendency then will be in the direction of over-production. There will be a surplus of food indicating that so much land in production is unnecessary, and the next year the amount of land cultivated will be reduced. This surplus may arise for various reasons. For example, a climate change may result in higher production, or some of the population may emigrate, lowering overall food requirements.

Thus we have a pair of countervailing tendencies. On the one hand is a tendency to have more children, or to attract migrants to help work the amount of land necessary to support the local population. This means the population will increase, requiring more land to be put into production and therefore more labor to work the extra land—a continual cycle. On the other hand is a tendency to take land out of production when a surplus of food is produced. The need for more field labor is thus reduced, and does not provide a particular incentive to have more children or attract migrants. It could even result in a decreasing local population if some force from outside the region was acting as a magnet for emmigration.

These countervailing tendencies can produce an increased population, a decreased population, or a stable one. The same tendencies may also result in increasing, decreasing, or stabilizing the amount of land in production. The point being that there is no necessary tendency for a population to grow, require an ever-increasing amount of food, and thus necessitate more land under cultivation. That may be the case under some conditions, but others may provoke exactly the opposite result, or even a stable population balanced between the desire for a larger population to help work the land, and a tendency towards a smaller population due to a deficit of production.

Once the full intensification of agriculture occurs (i.e. once there is no longer any fallow period), a variety of new tendencies begin to appear. Under a slash-and-burn system the easiest way to respond to social and biological events is either to create more fallow land, or to take more land out of fallow, giving rise to the sorts of changes described above. But what sorts of responses would be expected if all available land had already been taken out of fallow? The same pressures would likely be felt, but the solution of modifying the fallow period would be no longer available. At this juncture agriculture begins the modernization process.

Recall that burning the field was the major form of land preparation, the main purpose of which was to remove competitive vegetation— weeds. But much of the weedy material would have underground stems that would send up shoots and rapidly become a problem after burning. Thus, while burning certainly removes weeds, it does not do so completely. Probably in response to the persistent problem of weeds, the idea gradually developed of digging up the underground parts of the weeds, ushering in the era of the plow.

The plow would not have been of much use without draft animals to pull it, and likely for this reason, the plow never evolved in the Americas.[3] The plow as a major innovation generated other innovations, especially those involved with metallurgy. Thus the Bronze and Iron ages of the Old World never materialized in the Americas, possibly due to the lack of this all-important technological innovation. But apart from the role of agricultural modernization in shaping other aspects of world history, the plow initially led to the evolution of various agricultural implements, which made the farming enterprise ever more effective, and finally led to the complete modernization of agriculture. Before describing this system in detail, a very special form of agriculture, *enclave production*, must be considered. While marginal in the current modern world, enclave production is sometimes closely connected, both directly and indirectly, with the deforestation of tropical rain forests.

Agriculture and Enclave Production

There is nothing more British than tea time, yet that custom is an outgrowth not of ancient kings, dukes, and earls, but rather of British colonialism. Tea cannot be produced on the British Isles, and was not part of the culture until the colonies in South Asia were formed. Not trusting local people to take responsibility for anything as important as profits, the English East India Company evolved a form of production that proved to be extremely efficient, albeit at the expense of the people of India and the other colonies.

The key idea was to integrate all aspects of production, even very indirect ones. This enabled the British to maintain control over each component of production from planting the tea bushes, to harvesting the leaves, to drying and packaging the tea, and finally to shipping. Workers were both permanent and seasonal, but, as much as possible, all were housed in company-owned barracks and bought their provisions at company-owned stores. British employees hardly knew they were in India since all their needs were met on the company compound. It was as if there was a little Britain in the Indian highlands, which is why the system came to be known as enclave production. Furthermore, this sort of production can lead to major political influence, as witnessed by the central role played by the English East India Company in the history of Britain in South Asia.

The major remnant of this kind of production is banana production in Central and South America, which began quite differently than the tea enclaves of India, but whose consequences have been just as devastating for both the social and the physical environment of the region. Much as the British enclave production in Asia led to the growth and concentration of political power in the seventeenth and eighteenth century, so the international political scene in the American neocolonies of the United States was strongly influenced by the companies that controlled the enclave production of bananas. How was it that such power and influence arose?

Late in the nineteenth century, Lorenzo Dow Baker, a U.S. ship captain, picked up, on a stop in Jamaica, a small load of a curious yellow

fruit. Upon arrival in Boston he sold the load so rapidly that he immediately saw an immense future potential for this new agricultural product, and teamed up with Andrew Preston, an entrepreneurial friend, to
form the Boston Fruit Company. Baker purchased the fruit from producers in the Caribbean to bring to Boston, and Preston was in charge
of marketing in the U.S. The company flourished.[4]

About the same time, Minor C. Keith was looking to fulfill his destiny
as a railroad man. However, demand for new rail transport in the United
States had declined, and Keith faced the possible failure of the fulfill-
ment of his mission in life. He moved to Costa Rica where he obtained
a government concession to build a railroad from the central city of San
Jose to the port town of Limon. Having built the railroad he soon found
that Costa Rica's economy was not exactly dynamic, and there was really
very little for his new railroad to do. But just about that time he heard of
the remarkable success of the Boston Fruit Company and realized that
Costa Rica had a perfect climate for banana production. Keith saw that
bananas could give his railroad its *raison d'etre*, and got into the banana
business.[5]

The banana trade grew rapidly, with Guatemala and Honduras
joining Costa Rica as major production sites, and as many as ten companies producing or buying bananas and eight companies shipping
them. Then, in 1898, the Boston Fruit Company teamed up with
Keith's operation and purchased the major shipping and production
companies involved in the banana business, forming the United Fruit
Company. Guatemala, Honduras, and Costa Rica have not been the
same since.

The United Fruit Company grew by leaps and bounds over the next
half century. By the early 1950's it was the largest landowner in
Guatemala, and one of the largest in both Honduras and Costa Rica.
Unlike other foreign ventures that established businesses which inte-
grated with the rest of the national economy, the United Fruit Company
maintained the classic enclave production philosophy as invented by the
British. Every aspect of production remained under company control.
Along with its control was an ability to rapidly expand if conditions were

auspicious. United Fruit accumulated a great deal of land. Not only did it have a huge amount of land in production, it also held an equally large amount of land in reserve, possibly for speculative purposes, or perhaps for future banana production.

Export agriculturalists often purchase or even steal land situated on good soil from peasant farmers, forcing the farmers onto poorer soils on the hillsides, or in the rain forests. The United Fruit Company in Guatemala is an excellent example.6 By 1950 United Fruit owned 565,000 acres of land, making it the largest landholder in Guatemala. At the same time 75% of the peasant families were either without land at all, or had small plots of land that were marginal at best. Such an arrangement was clearly tinder for a social explosion.

The social explosion came, not as a violent outburst, but rather as a sophisticated and civilized political campaign. In 1950 one presidential candidate, Jacobo Arbenz, ran on a ticket of agrarian reform. He promised to purchase land held by large landowners and redistribute it to small peasant farmers. Arbenz won the election on that platform and his agrarian reform program was rapidly put into action. On farms of more than 90 hectares (223 acres), land not being used for production was targeted for expropriation. The expropriations were made up of lands taken from 1,059 farms with an average size of 4300 acres. Approximately 100,000 peasants received title to the land thus expropriated. It is not difficult to see why Arbenz was popular.

Probably Arbenz's undoing was the expropriation of United Fruit Company land. While the United Fruit Company was the largest landholder in Guatemala, only 9% of its land was actually being used. Consequently about 240,000 acres on the Pacific Coast and 173,000 acres on the Atlantic coast were expropriated, for which the company was compensated with $6,000,000, based on its own stated value of the land.

Arbenz had challenged the basic arrangement of the Guatemalan economy.7 To the politicians and their clients in the United States of America, such a challenge seemed a harbinger of things to come and caused a great deal of consternation. If it was true, as the president of

Castle and Cook (parent company of United Fruit Company) put it,[8] that "Without overseas investment your entire economic infrastructure collapses in on you," Guatemala's taking charge of its own resources was certainly a threat to free-wheeling investment opportunities available in the Third World. And if Guatemala, as one of several dozen countries that performed the same function, was not particularly important in and of itself, what might happen in Costa Rica or Nicaragua or Honduras if Guatemala were successful? Indeed, shortly after Arbenz's victory and initial confrontations with the United Fruit Company, both Costa Rica and Honduras began making demands of the United Fruit Company, presumably buoyed by Arbenz's success.

Additionally, the United Fruit Company was a powerful entity not, only in Guatemala, but in the United States. As the owner of over 300 million acres of land, 2000 miles of railroads, and 100 steamships in Central America, the United Fruit Company was an influential player in international politics. Furthermore, Secretary of State Dulles was a senior partner in the Sullivan and Cromwell law firm, the principal legal agent for the United Fruit Company. His brother was, of course, the head of the U.S. Central Intelligence Agency (CIA).

Finally, and most importantly, at about this time, the Cold War was becoming a defining feature of U.S. politics. The Eisenhower administration (especially Vice-President Nixon) was to set the stage for the cold warriors.[9] The international communist movement advocated quite a theoretical package for the oppressed, and it was not difficult to see communists behind every threat against a capitalist venture wherever it occurred. Thus, despite the evident non-communist nature of Arbenz himself, the fact that his legislature contained several deputies who were members of the communist party, and the simple fact that he promoted agrarian reform, either led the cold warriors to believe there was a communist conspiracy afoot, or at least enabled them to justify their actions based on the claim of a communist conspiracy.

In 1954 the Eisenhower administration went into action. A lesser colonel in the Guatemalan army, Castillo Armas, was chosen by the

CIA to "lead" the revolution against the Guatemalan government. The plan had been hatched almost a year earlier at the upper levels of the U.S. government. The CIA knew that neither the Guatemalan people nor the Guatemalan military would rally to the cause of overthrowing the popular Arbenz, so they had to rely on a strategy of causing Arbenz to surrender without a true military threat. This meant that they had to create the impression of a real threat to convince Arbenz that he had no way of avoiding a bloodbath other than resigning. Through some duplicitous diplomacy and a great deal of propaganda, the Arbenz government was led to believe that Castillo Armas had a large force (he never actually had more than 300 troops) poised to engage the Guatemalan Army, and that U.S. armed forces were ready to enter the fray in support of the overthrow. Most spectacular was the bombing of Guatemala City by CIA planes, generating panic, as planned, in the city. Arbenz, fearing a bloodbath, capitulated, following almost exactly the script the CIA had plotted.[10]

Today, despite retaining its basic enclave form, the modern mode of banana production is technically far more sophisticated than it was in the days of Arbenz. To establish a modern banana plantation it is often necessary to construct a complex system of hydrological control wherein the soil is leveled and crisscrossed with drainage channels. This significantly alters the physical nature of the soil. Contemporary banana production includes burying plastic tubing in the ground to eliminate the natural variability in subsurface water depth. Metal monorails hang from braces placed into cement footings to haul the bunches of bananas (Figure 4.1). To avert fungal diseases, heavy use of fungicides is required, and because of the large scale of the operation (Figure 4.2) chemical methods of pest control are the preferred option. The banana plants create an almost complete shade cover and thus replace all residual vegetation. Pesticide application is sometimes intense, other times almost absent, depending on conditions, but over the long run one can expect an enormous cumulative input of pesticides, the long-term consequences of which are unknown but unlikely to be salutary.

Figure 4.1. Banana production in Central America. Photo on top shows the metal monorail that runs throughout the plantations. Photo on bottom shows banana workers harvesting a bunch of bananas. Photos are from the Standard Fruit Company plantations in Rio Frio, Costa Rica.

Figure 4.2. Aerial view of United Fruit Company banana plantation in Costa Rica.

Figure 4.3. Banana workers prepare metal monorail to receive harvested bananas to bring them to the packing plant.

A major social transformation is also required to set up a modern banana plantation. The factory-like conditions of production require a semi-proletarianized work force, sometimes drawn from the local peasant population, but usually brought in from surrounding areas (figure 4.3). Banana production tends to promote a local overpopulation crisis by encouraging a great deal of migration into the area in which it operates. As the international market for bananas ebbs and flows, workers are alternatively hired and fired. When fired, there is little alternative economic opportunity in banana zones, and displaced workers must either look for a piece of land to farm, or migrate to the cities to join the swelling ranks of shanty-town dwellers.

The Modernization of Agriculture in the First World

Understanding agriculture in a country like the United States is sometimes hindered by an overly romantic notion of the farm. As part of the original expansion of European America and something of a centerpiece of democracy "American style," the rugged farmer carving a homestead out of the wilderness is an important piece of lore. While the small family farm may have been important in both the Europeanization of America, and the establishment of the form of democracy practiced earlier in our history, the farm and farmer of today bear little resemblance to this romantic vision. Such romanticism is fueled by a confusion between *farming* and *agriculture.*

Farming is a resource-transformation process in which land, seed, and labor, are converted into peanuts, for example. It is Farmer Brown cultivating the land, sowing the seed, and harvesting the peanuts. Agriculture is the decision to invest money in this year's peanut production; the use of a tractor and cultivator to prepare the land; an automatic seeder for planting; application of herbicides, insecticides, fungicides, nematocides, and bacteriocides to kill unwanted pieces of the ecology; automatic harvest of the commodity; sale of the commodity to a processing company where it is ground up and emulsifiers, taste enhancers, stabilizers, and preservatives are added; packing in convenient, pleasing-to-the-consumer jars; and finally marketing under the trade name of

Skippy or Dippy. In short, while farming is the production of peanuts from the land, agriculture is the production of peanut butter from petroleum.[11] Over the last two hundred years, and especially in the last fifty, farming has been transformed into agriculture. This transformation has included impressive technical as well as socioeconomic changes.

The Technical Side of Modern Agriculture

The transformation of farming into agriculture developed in two recognizable waves. The first wave commenced at the end of the Civil War and culminated earlier in this century with the emergence of the hydrocarbon society. The second wave started towards the end of World War II and is still evolving.

Before the outbreak of the U.S. Civil War, Cyrus McCormic built a revolutionary machine, a device you could attach to a team of horses and pull through the field to harvest automatically. Surprisingly, McCormic's automatic harvester did not come into use until well after the Civil War had begun, despite the fact that it was well-known some twenty years earlier. And it was not cynicism farmers had about "newfangled devices," it was a rational assessment of economic facts that caused them to retain a technology that required five people per acre at harvest time, when they were well-aware of this new technology that required only one.

The all-important factor was availability of labor. At the time the U.S. Civil War broke out, the westward expansion of agriculture, which had populated the Ohio valley and was in the process of populating lands further west, required a supply of harvest labor easily met by the comparatively large population of potential workers.[12] The many men (very few women) looking to establish homesteads provided a ready source of cheap harvest labor. With all that cheap labor available, it made no sense to invest a large amount of money in an automatic harvester, and the McCormic reaper remained an underutilized "newfangled" device for nearly two decades. Thus, McCormic's reaper, which eventually revolutionized agriculture and strongly contributed to the Industrial Revolution in the United States, was ignored for two decades, not because of unreasonable or backward farmers, but because of logical economic calculations.

With the outbreak of the Civil War, labor grew short, and the McCormic reaper became widely adopted, initiating a change in the mentality of farmers. "Factories in the fields,"[13] was to become the future of agriculture, and a host of automation devices from corn harvesters to threshers to automatic seeders and more, followed McCormic's reaper. At this point power was a limiting factor. The horse team was replaced by the steam engine for certain processes, but many processes could not conveniently utilize steam engines.

The next great event within the first technological wave occurred in the early part of the present century. The internal combustion engine made automated devices even more efficient. The introduction of the internal combustion engine, effectively the introduction of fossil fuel-based traction power, completed the first major transformation of farming to agriculture.

Finally, in conjunction with chemical research aimed at the development of weapons and the control of tropical diseases during World War II, a miracle seemed to emerge. The miracle was a chemical, dichloro-diphenyl-trichloro-ethane, DDT. Pests could be killed in the field much like enemies could be killed in war, and indeed, the martial metaphor was skillfully used to transform American agriculture.[14] In addition to this miracle, the munitions industry, which blossomed as a result of research into new explosives based largely on nitrate chemistry, gave rise to technology for producing and using chemical fertilizers in agriculture.

Agriculture soon became fully dependent on chemicals, pesticides, and fertilizers, most of which were formulated from petroleum. This completed the technological transformation of agriculture and set up the foundations of "modern" agriculture. Today a farm is a technologically sophisticated factory that uses fossil fuels to drive machinery which transforms the soil physically and chemically; uses products derived from fossil fuels to eliminate unwanted components of the ecology (i.e. pests); and adds other components that were either not there to start with, or had become exhausted from prior use (i.e. nutrients).

The Socioeconomic Side of Modern Agriculture

In conjunction with this massive technological transformation, the socioeconomic structure of agricultural production underwent an equivalent major structural change. This transformation had to do with the way in which critical agricultural commodities were bought and sold. As far back in history as one cares to trace, it has been the norm that agricultural products were traded. Indeed intermediaries were often involved. But at the close of the eighteenth century a food crisis loomed in Western Europe. Wheat had always been a staple crop, but the form in which it was eaten was class-dependent. The lower classes ate mainly porridge or gruel, soup-like preparations that required little grain to fill a stomach. The upper classes ate a more elegant product, bread. To make bread the wheat had to be ground into flour, mixed with water and yeast, kneaded into dough, allowed to rise, and finally baked. Such a large input of labor could be afforded only by the well-to-do. The peasant class continued to have to be satisfied with its porridge.[15]

During this period the long-distance trade in agricultural products was largely restricted to luxury goods such a chocolate, tea, coffee, and sugar. The main agricultural goods that fed the Industrial Revolution, cotton and wheat, were hardly traded internationally before the late eighteenth century. There was simply not enough demand since cloth was made mainly from wool and local production of wheat for porridge and bread fully satisfied all needs. But towards the end of the eighteenth century two factors combined to dramatically change this pattern. The first factor was the evolution of improved technology to mill grain. The second was a change in the social organization of work.

With the growing industrialization of Western Europe, the dominance of porridge and gruels began to decline and the working class began to eat bread. Improved milling technology had caused the price of bread to decline, making it more available to the masses. More important, it was difficult for factory workers to bring gruel or porridge to work in their lunch buckets. The result was a dramatic increase in bread consumption and a concomitant increase in the demand for wheat. In almost all European countries local agricultural production simply could not keep pace. This led to a dramatic increase in the international trade in wheat which, in turn, led to the first important international wheat-trading business complex the world had ever known. It included key points at Liverpool, Constantinople, and Odessa.

As part of the consolidation of the Russian empire in the sixteenth century, colonists were given land in exchange for settling in southern Russia (what is today the Ukraine and Moldova). These colonists were given some of the finest agricultural real estate in the world, on par with the rich soils of the U.S. Great Plains and the Pampas of Argentina. Their product was sold at the most natural market, southern Russia and the entrance to the Black Sea, the town of Odessa. Odessa became the market for the hundreds of thousands of farms that dotted the surrounding countryside.

At that time the grain trade was a chaotic business. Grain was purchased directly, with little predictability about when, or how much, grain would be available to millers, bakers, and thence the general populace. This was especially critical in industrializing countries, as the growing work force was becoming increasingly dependent on bread. A shortage of this critical resource had the potential to create a great deal of social instability. Soon clever grain merchants, realizing that overland travel of a courier was far quicker than sea travel of a lumbering merchant ship, developed a system in which they took samples of grain from ships as they sailed through the narrow straits at Constantinople, carried them overland to the trading houses of Europe, and sold in-transit shipments based on samples. This was the beginning of *futures trading*, a financial speculative strategy that would become extremely important to the industrial capitalist system.

Agreements to buy and sell grain shipments were usually made in clubs or informal gathering places, such as coffee houses. Liverpool became the center for buying grain destined for the mills of England. The wheeler-dealers of the time knew where to meet with one another. Gradually these centers became formalized and eventually evolved into the giant commodity trading centers we know today.

Thus, by the end of the eighteenth century in industrially burgeoning Europe, we had the beginnings of a complex grain trade, which involved 1) a serf farmer who sold grain (or gave tribute) to 2) a landowner (usually nobility in Russia) who sold it to 3) a transporter who transported it to Odessa where 4) a shipper bought it and sailed to Constantinople where 5) a representative of the shipper took a sample and brought it overland to 6) a primitive commodity exchange center in Liverpool or another European city where 7) a local merchant bought it to be sold to 8) a miller. Into this system stepped several entrepreneurial types. A typical example was Leopold Louis-Dreyfus.

At the age of seventeen Louis-Dreyfus left his family's farm and sold his share of the season's wheat harvest. At the time, commerce in Russian wheat was in the doldrums due to the expansion of other international sources of wheat, and to the dominance of Greek shipping interests in the Black Sea. Louis-Dreyfus spent several years in the stream of wandering merchants involved in the European grain trade. He amassed a considerable amount of capital in grain trading, established his own grain company, and took out large loans aimed at enlarging his enterprise. With his firm and finances in place, Louis-Dreyfus arrived in Odessa in the 1860's and purchased the bulk of the grain storage bins in the city. He then sent out his own agents, not to buy grain from farmers, but to sign contracts to purchase their grain in the future. At the other end of his enterprise, in Western Europe, he sold those contracts for future delivery. By the 1870's he was contracting for wheat from the Russian hinterland, shipping by rail to his storage facilities in Odessa, shipping the grain on his freighters, and selling futures and grain to buyers in Hamburg, Bremen, Berlin, Mannheim, Duisburg, and Paris. The Louis-Dreyfus Company had become the first giant grain company in the world.

Similar stories can be told for four other giants, Bunge of Argentina, Continental of New York, Cargill of Minneapolis, and André of Switzerland. With Louis-Dreyfus, these five giants dominate the world grain trade much like the Seven Sisters[16] dominate the world petroleum trade. For all practical purposes, all the grain produced in the developed world must be sold to one of these giants.

While this pattern is most highly developed in the grain trade, similar structures developed in other sectors of the food industry with
canned vegetables, processed foods, and so on. The story, beginning
with the Industrial Revolution and continuing with only minor varia-
tion today, is a tendency for buyers of agricultural products to become
large, and the tendency for most financial decisions to be made far from
their source, the farmer. By the time Louis-Dreyfus arrived in Odessa,
the serf farmers of Russia had little knowledge of, or control over, what
happened to their grain after it left their farms. But those who purchased
it were involved in high finance throughout the industrial world.

The Modern System

The above history sets the framework for understanding the modern
agricultural system. On the one hand, technological innovation has created a massive need for modern inputs into the agricultural system,
which the farmer is forced to buy. On the other hand, socioeconomic
"innovation" has created a massive marketing and processing structure,
to which the farmer must sell. Enclave production remains something of
an anachronism in this modern system, and would not even be mentioned were it not so important, as we shall see, in areas experiencing
rapid tropical rain forest deforestation.

---

[1] This point is discussed in detail in Chapter 6. Also see Westoby, 1989.

[2] This analysis is based on the pioneering work of Boserup, 1965. The specific presentation here is after Minc and Vandermeer, 1990.

[3] In the Americas, potential draft animals, such as horses or oxen, had gone
extinct earlier.

[4] McCann, 1976.

[5] ibid.

[6] A great deal has been written on the role of the United Fruit Company in
Guatemala. Key references include, Melville and Melville, 1971;
Immerman 1982; Rabe, 1988.

[7] This is a crucial conceptual issue that will be discussed in detail in the next
chapter.

[8] Quoted in the film *Controlling Interest*.

[9] Rabe, 1988.

[10] There has been a great deal of speculation as to why the CIA was dis-

patched to overthrow a democratically-elected government. The official line is that the Arbenz government was a communist one, or at least very sympathetic to communism, and therefore could not be allowed to prosper. In the hysteria of the times, this could very well have been true. But any sensible analysis of the Arbenz program would have easily demonstrated that its program was far from what Moscow might have wanted. A more cynical interpretation suggests that the Dulles brothers, with their important connections to the Sullivan and Cromwell law firm, were simply protecting their own investments in the United Fruit Company. This is the impression one gets from McCann (1976). The truth is probably a little bit of both.

11 This is the imagery first provided by Lewontin, 1982.

12 Hired hands and plowboys.

13 In her famous and remarkably insightful analysis of the general tendencies of modern agriculture, Carey McWilliams warned of the immense problems we faced because of the tendencies of modern agriculture. *Factories in the Field* (McWilliams, 1939) was the first convincing critique of modern agricultural production, calling for its transformation well before the development of a serious alternative agriculture movement.

14 The purposeful development of this metaphor as a marketing tool is described in detail by Russell, 1995.

15 Morgan, 1979.

16 Since Chevron purchased Gulf in 1984, the famous seven sisters became six — Chevron, British Petroleum, Royal Dutch/Shell, Exxon, Mobil, and Texaco.

5

◆

The Multiple Faces of Agriculture in the Modern World System

Much has been written about the emergence of the current world order, and as a product of the past five centuries of European expansion, the clear split between the First and Third Worlds. It is hardly debatable that the changes induced by this expansion were among the most spectacular in the history of humanity. The structure of today's world was established at the time European merchant societies took control of world trade from the Chinese/Arab monopoly. Europe used the vehicle of international trade and the incorporation of colonial outposts to supply raw materials to fuel their industrial development. For the most part these raw materials were either luxury products (e.g. sugar, spices); drugs (e.g. coffee, tea); or industrial inputs (e.g. wood, cotton, indigo). Most frequently they were agricultural products.

The Third World evolved from those early colonies, and industrial capitalism became the dominant form of production in the developed world. Yet the world came to operate as a single economic entity. The notion of industrial capitalism existing in the First World and some more primitive form of capitalism existing in the Third World is outmoded. The Third World is not simply "waiting" for the sort of economic development that made the First World what it is. Rather our globe is one world system, connected in a complicated network. To understand how and why rain forests are disappearing, it is necessary to understand that network.

The Modern World System

The notion that the world is intricately connected, that it no longer makes sense to try understanding isolated pockets, such as nations, is most commonly associated with the sociologist Immanuel Wallerstein.[1] We agree with his general assessment, and add that isolated thematic pockets are similarly incomprehensible unless embedded in this global framework. For this reason, attempts at understanding tropical rain forest destruction in isolation have largely failed. As should be clear by now, the fate of the rain forest is intimately tied to various agricultural activities. These agricultural activities are embedded in larger structures, some retaining a connection to agriculture, some not.

In this chapter we outline the components of the modern system that impact on the phenomenon of deforestation. First, we consider the operation of Developed World agriculture, and note its special form, distinct from that which dominated only a century ago. Second, we look more generally at the dynamics of Developed World economies. Finally, we analyze Third World economics as related to Developed World structures.

Agriculture in the Developed World Today[2]

As described in the previous chapter, historical developments gave rise to a particular model we refer to as the modern agricultural system. In modern agriculture we have three units, the suppliers who supply the inputs to the farmer, the unit that is the farm itself (farmers), and the unit to which the farmer supplies the products. Immediately apparent are problems stemming from the very nature of this structural arrangement.

In figure 5.1 we illustrate this relationship, showing the transfer of materials (figure 5.1a), the transfer of money (figure 5.1b, horizontal arrows), and the rational economic desires of all three components (figure 5.1b, vertical arrows). The suppliers obviously wish to get the most money possible for their supplies, while the farmer wants to pay as little as possible for them. The buyers want to pay as little as possible for the commodities they purchase from the farmer, while the farmer wishes them to pay as much as possible. There are inevitable tensions here. If we really had that textbook fantasy, a world of perfect competition and free trade, one might expect that these tensions would resolve themselves to produce fair prices and terms of exchange among the units involved in this three-part process. But in the real world various forms of political power alter that theoretical free market.

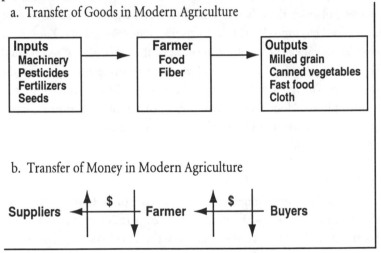

a. Transfer of Goods in Modern Agriculture

b. Transfer of Money in Modern Agriculture

Figure 5.1. Diagram of goods (a) and money (b) transfers in the modern agricultural system.

The average U.S. farmer of the post-World War II era was a small entrepreneur running the so-called family farm. Granted such farmers had long before made the transition from only producing food for family consumption, or to exchange for other goods, to a more business-like stance of investing money to make a profit. Yet the scale of operation remained exceedingly small when compared with suppliers like Dow Chemical or John Deere, giant corporations with monopoly-like control over the markets in which they participate, or equivalent-sized buyers like Skippy Peanut Butter or the Cargill Grain Company. It does not take a Ph.D. in economics to predict what is likely to happen given such an arrangement. Farmers, with little economic leverage because of their isolated condition and relatively small size, were constantly squeezed on both ends of the system. Suppliers extorted large sums of money for necessary inputs, and buyers refused to pay fair prices for products (at least that is the way the farmer tends to see it), squeezing the poor farmer in the middle. This is the fundamental structure of agriculture in the U.S. and much of the rest of the Developed World.

The implications of this modern structure are many and diverse, but for our purposes there is one critical feature. Agriculture has become part of the Developed World's industrial system. While it may have made sense in the nineteenth century to speak of the "agricultural sector," and the "industrial sector," and to analyze them separately, such is not the case today. The independent farmer has become, in a sense, a worker in the metaphorical food factory that is modern agriculture. As such, developed world agriculture today follows the same rules as the rest of the industrial system.

Developed World Economics[3]

Given today's interconnected world, in order to understand the Third World we must view it with reference to the operation of the First World (the developed countries). We must view the Third World as embedded in the modern industrial system. In that system the people who provide the labor in the production processes are not the same people who provide the tools, machines, and factories. The former are the workers in the factories, the latter are the owners of the factories. The owners, who own the machines and tools, directly make the management decisions about all production processes. A good manager tries to minimize all production costs, including the cost of labor.

However, the owners of the factories face a complicated and contradictory task. While factory workers constitute a cost of production to be minimized, they also participate, along with the multitudes of other workers in society, in the consumption of products. In trying to maximize profits, factory owners are concerned that their factories' products sell for a high price. This can only happen if workers, in general, are making a lot of money. In contradistinction to what is desired at the level of the factory (for the workers to make as little as possible), at the level of society the opposite goal is sought (for the workers to make as much as possible). Factory owners must wear two hats then, one as owners of the factories, and another as members of a social class. As the owner, he or she wishes the laborers to receive as little as possible, but as a member of the social class, he or she benefits if laborers in general receive as much as possible (to enable them to purchase the products produced in the factory).

As factories compete with one another, labor, being one of the major costs of production, is always a target for reduction. Competition forces individual factories to mechanize processes whenever possible, giving them a competitive edge, at least temporarily, over those factories which have not mechanized. Unfortunately, mechanization means reduced employment, which means lowered general consumption power,[4] which means economic recession. But economic recession means a reserve of laborers ready to take up new employment, at low wages, which stimulates the opening of new economic sectors, and economic recovery. Since the new economic sectors are comprised of individual competing factories, the whole process starts over.

Phases of economic expansion and contraction are not desirable. Economic contraction is classically associated with political instability, something that factory owners and their political representatives naturally wish to avoid. Furthermore, and probably more importantly, economic contraction means profits are lowered and generally more difficult to come by. Lowered profits mean less money for investment and thus less chance of opening up new economic sectors to generate an economic revival. If new economic sectors do not appear, the general process of economic contraction will continue unabated.

Thus factory owners (as well as the general population) view such economic contraction as a crisis. On the one hand, it is a crisis at a general level in that it can lead to the demise of the entire economic system. On the other hand, and more importantly for our purposes here, it is a crisis for individual factory owners in that their ultimate goal, to make a profit on investment, is threatened. Just how such crises are resolved is an important part of the history of the development of the industrial capitalist system, a point to which we return below.

Third World Economics: Dependency Theory

As difficult as life may get for some citizens of the United States or other developed nations, one can hardly fail to note a dramatic difference from conditions of life in the Third World — in the so-called developing nations. It is in the Third World that agricultural production has seemingly not been able to keep up with population growth. It is also in the Third World that dangerous production processes are located, where raw materials and labor are supplied to many industries at ridiculously low cost, where people live in desperation, and where talk of bettering the state of the environment is frequently met with astonishment. "How can you expect me to worry about deforestation when I must spend all my worry time on where I will find the next meal for my children and myself?"

Late nineteenth and early twentieth century analysts generally agreed on the basic mechanism causing such a pattern. It was thought the Third World was simply behind, and given time, Third World nations would develop, just as the developed countries had. This "time" hypothesis was the received wisdom for a long period. Yet, by the end of World War II it had become evident that not only was the Third World not developing, the trend was actually the reverse. Things were getting worse not better.

This simple observation led to a great deal of analysis, beginning in the 1950's, but with the most vigorous discussion and debate occurring during the 1960's and 1970's. The general field of analysis is now known as Dependency Theory. In its most general form, Dependency Theory holds that the underdevelopment of the Third World is not an accident of history or a product of bad real estate, but an organic outgrowth of the progression of the Developed World. Underdevelopment in the

Third World is seen as an outgrowth of events in the First World, and as somehow necessary for the maintenance of the First World. Many dependency theorists have elaborated on this theme, and hundreds of books have been written, representing various positions within the general idea of dependency.[5] The one, and we think only, thing on which they agree, is that the underdevelopment of the Third World is a consequence of the development of the First World.

To summarize the complex and historically conditioned debates within this vigorous field of political economy would require another entire book. Rather let us summarize what appears to be the most recent and lucid explanation of the phenomenon, one that more-or-less summarizes what others have said, and condenses the arguments into a series of simple ideas.[6]

Recall the heuristic characterization of the industrial system as consisting of two classes of people, the owners and the workers. The owners had to wear two hats in making decisions — wanting to pay their workers as little as possible, but wishing for workers in general to make as much as possible. This contradictory position represents the "engine" that drives economic growth in a modern capitalist economy. Production processes are mechanized to reduce labor costs, making more workers available at lower labor costs, so that new economic activities are then able to gain a foothold.

The situation in much of the Third World appears superficially similar. For the most part we are dealing with agrarian economies in which there are two obvious social classes, those who produce crops for export like cotton, coffee, tea, rubber, bananas, chocolate, beef, and sugar; and those who produce food for their own consumption on their own small farms and, when necessary, provide the labor for export-crop producers.

Consider, for example, Carlos, a Costa Rican farmer we used to know. He worked for the United Fruit Company for over ten years. Because of a temporary drop in the price of bananas on the international market, the company decided to scale back production. Carlos lost his job. United Fruit was the only employer in the area, so Carlos and his family *optime for workers:*

unions

were forced to either migrate to another area where a job might be found, or to find a piece of land and carve out a farm. They chose the latter.

The productive part of their farm is approximately two hectares on which they grow cassava, taro, and several small fruit trees which are not yet fruit bearing. They also have chickens, and a cow. Carlos feeds his family on the root crops, the eggs from his chickens, and the milk from his cow, and purchases some basic grains with the small amount of cassava he can sell. The last time we saw Carlos, a German company had started an ornamental plant farm up the road, and he had gotten a job there as a night watchman. He told us, "at least I won't forget what money looks like." His salary was about $3.00 a day.

Another acquaintance, Bruce (not his real name), was Chief Executive of an international corporation involved in export-crop production. After we sat in his living room listening to him close a million dollar deal over the telephone, he recounted how his company was relocating much of its operation to Honduras where the military, "knows how to control the unions," unlike Costa Rica, where strong democratic traditions maintain the rights of workers to organize.[7] However, his main concern was with the threat of a U.S. recession, since his company exported luxury products there. That the Costa Rican unions would have made wages higher for Costa Rican workers and would eventually promote the growth of the economy was obviously of no concern to him (except of course in the negative sense that his company would have had to pay them more).

Superficially these people appear similar to the factory owner and worker of the First World system. Yes, Bruce runs a "factory" and Carlos works (or worked) in a "factory." However, Bruce is not concerned with selling commodities from his production process to Carlos and people like him, but rather to consumers in the First World. Third World export producers do not wear two hats with regard to the laborers on the farm. Their concern is not to sell products to those workers, but rather to the workers in the First World. What this means is that the dynamo of economic growth created by the contradictory goals of the factory owner in the First World simply does not exist in the agrarian Third World. The First World factory owner is concerned with the general economic health of the First World working class, and thus there is social pressure to maintain consumption power in that class. There is general agreement that while such an arrangement tends to produce economic cycles of boom and bust, it also represents the mechanism of economic growth, and is therefore the base of development. But that base of development does not exist, to a significant extent, in the Third World. Bruce could care less whether Carlos can buy cotton or bananas or sugar. Bruce's concern is whether Detroit auto workers and San Francisco yuppies can buy his products.

In figure 5.2 the idea is represented diagrammatically, assuming the First World country produces only automobiles and blue jeans and the Third World country produces cotton and bananas. Just follow the arrows. For example, the auto workers and textile workers "pay money for autos to" the owner of the auto factory who "pays money for labor to" the auto workers. We hope it is obvious from the diagram that in the Developed World the money that goes to labor eventually goes to purchase the products, thus providing the machinery of economic growth. The Third World lacks an arrow connecting the export producers (factory owners) to the traditional farmers/workers in the process of consumption. The workers and factory owners of the Developed World are *not* the consumers of Third World products.

The situation in the First World as pictured in figure 5.2 may be described as *articulated*. Articulation is used here in the anatomical sense of having joints or articulations that are composed of connected segments. The typical arrangement in the Developed World is an articulated economy, while that in the Third World is disarticulated, in that the two main sectors of the economy are not articulated, connected, with one another. Furthermore, since the production systems of the traditional agriculturalists and the export agriculturalists are not connected, the economy is also referred to as *dual* — the traditional sector operates relatively independently from the export sector.

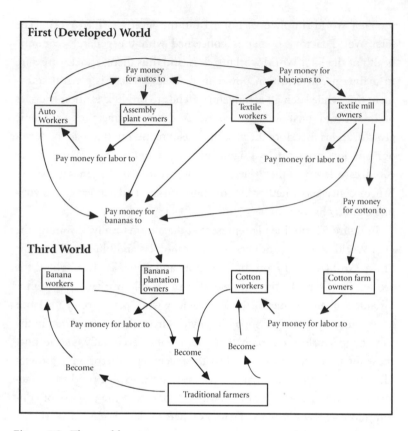

Figure 5.2. The world system.

This disarticulation, or dualism, goes a long way towards explaining the differences between analogous classes in the First and Third Worlds. Flower producers in Colombia do not concern themselves much over the fact that their workers cannot buy their products. On the other hand, the factory owners in the U.S., whether they manage private factories or run government owned and/or subsidized industries, care quite a lot that the working class has purchasing power. General Motors "cares" that the general population in the U.S. can afford to buy cars. Naturally they aim to pay their own workers as little as possible, but that goal is balanced by their wish for the workers in general to be good consumers.

So the different economic structures—articulated in the First World, disarticulated or dual in the Third World—go a long way towards explaining the relative conditions of the lower social classes in those areas of the world. What remains to be explained, and this is a more difficult concept, is how this dualism in the Third World is maintained.

The Function of the Dual Third-World Economy

The climate for investment is variable in the Developed World. There are times when it is difficult to find profitable investments at home. At such times it is useful to have alternatives to investment. The Third World provides those opportunities. The Developed World, because of its basic structure, tends to go through cycles of bust and boom, sometimes severe, other times merely annoying. During low economic times, where is an investor supposed to invest? Clearly without the presence of a Third World, the crisis would be of a qualitatively different sort, in that the Third World provides a sink for investments during rough times in the First World. This is why the dualism of the Third World is a *functional dualism*. It functions to provide an escape valve for investors from the Developed World. The West German entrepreneur who started the ornamental plant farm on which Carlos worked as a night watchman, invested his money in Costa Rica because opportunities in his native Germany were scarce at the time. What would he have done had there been no Carlos willing to work for practically nothing, and no Costa Rica willing to accept his investments at very low taxation? Clearly Costa Rica is, for him, a place to make his capital work until the situation clears up in Germany. Union Carbide located its plant in Bhopal, India, and not in Grand Rapids, Michigan. U.S. pesticide companies export insecticides which have been banned at home, to Third World countries. U.S. pharmaceutical companies pollute the ground water in Puerto Rico because they cannot do so (at least not so easily) in the United States. In all cases, Third World peoples are forced to accept such arrangements, largely because of their extremely underdeveloped economies.

With this analysis, the origin of the Third World as an outgrowth of European expansion (while a correct and useful historical point of view), can be seen as not the only factor to be considered. Even today, the maintenance of the Third World is a consequence of the way our world system operates. The Developed World remains successful at economic development for two reasons. First, because it has an articulated economy, and second, because it is able to weather the storm of economic crisis by seeking investment opportunities in the Third World. The Third World in contrast has been so unsuccessful because its economy is disarticulated, lacking the connections that would make it grow in the same way as the Developed World. Yet at a more macro scale, the dualism of the Third World is quite functional, in that it maintains the situation in which investors from the Developed World can use the Third World as an escape valve in times of crisis. Indeed, it appears that the Developed World remains developed, at least in part, specifically because the Underdeveloped World is underdeveloped. This is the point of the dependency theorists.

This picture is very general and certainly does not strictly apply across the board. Many countries do not fit this model. For example Taiwan and South Korea are Third World countries that experienced remarkable levels of economic growth; Hong Kong and Singapore have few peasant farmers; in Finland capitalists have few investments in the Third World; and none of the command economies (the eastern European countries) had significant investments in the Third World. Each of these examples has particular exceptional circumstances associated with it. Far more common are El Salvador, Nicaragua, Costa Rica, Colombia, the Philippines, Thailand, Vietnam, Mozambique, Zaire, and so forth.

The Modern World System and Tropical Deforestation

Given such a socioeconomic structure, it is easy to see how tropical rain forests become targets for deforestation. At times of low market value for tropical export crops, rural proletarians lose their jobs and are forced to return to the countryside and a life of farming. Since most of the existing agricultural land is devoted to those very export crops, the former proletarian is forced to look elsewhere for land to farm. Frequently the rain forest is the only land available.

Complicating this picture is the position of the logger, as described in the next chapter. Cutting logs out of a tropical forest does not necessarily represent deforestation in and of itself. However, in the context of a Third World disarticulated economy, when logs are cut from a forest, landless peasants are nearly always waiting to follow the logging roads and take advantage of land that has been at least partially cleared of trees.

As we have already emphasized, lack of land and thus lack of food security, is the driving force of deforestation. The logged forest could easily regenerate into a tropical rain forest, were it not for the additional conversion to agriculture. So the force that creates landless peasants must be identified as a major force in the destruction of tropical rain forests. That force is the modern world system.

Yet there is a further complication to all of this. Export agriculture itself is also implicated in tropical rain forest destruction. Malaysian rain forests are directly cleared to make room for rubber plantations. Forests in Costa Rica are directly cleared by banana companies. While this direct form of conversion to modern agriculture is evident and spectacular in several pockets (e.g. Brazil, Malaysia, Indonesia), it is probably not the most common form of rain forest destruction. When direct conversion occurs, it is probably more damaging than peasant agriculture, but its impact world-wide is less.[8]

However, as we have indicated in this chapter, the secondary impact of export agriculture is dramatic. An additional component of this secondary impact is the need for Third World governments to continue expanding this export agriculture—likewise an inevitable consequence of the underlying structure of the general world system.

We thus come to what we regard as a disturbing conclusion. An analysis of the socioeconomic aspects of current conditions in the world, especially as they relate to areas where most of the world's rain forests remain, suggests that the basic structure of that world system is what maintains the pattern of land and food insecurity, which drives

small-scale agriculture into rain forest areas. Saving the world's rain
forests thus requires transforming that world system.

---

[1] Wallerstein, 1980.

[2] This section follows closely the analysis of Lewontin, 1982.

[3] This, of course, is a topic for a textbook. We do not pretend to summarize
economics as an academic discipline, but rather hope to describe those
components important in understanding how the Third World functions.
Since our goal is to comprehend the loss of tropical rain forests, and almost
all of them are located in the Third World, we need to understand Third
World issues. Part of grasping Third World issues is understanding how the
Developed World impacts upon them. That is what this section seeks to
articulate. The basic framework has been described many times before,
classically through Ricardo or Marx, or more recently, even in standard
textbooks (e.g. Samuelson, 1970).

[4] This is what is usually referred to as an underconsumption crisis (Haberler,
1958)

[5] For different perspectives see: Kanth, 1994.

[6] de Janvry, 1982.

[7] This was in the late 1970's, before the spread of Solidarismo, an anti-union
movement that has more recently clamped a strong hold on labor organizing in Costa Rica.

[8] For example, shifting cultivation (slash-and-burn agriculture) practices in
Africa account for 70% of the clearing of closed-canopy forest (Brown and
Thomas, 1990). Estimates of the number of farmers engaged in clearing
forest lands in the humid tropics (including primary and secondary
forests) each year range from 300 million to 500 million. Assessments of
the area cleared by peasant farmers range from 7 million to 20 million
hectares each year (National Research Council, 1993).

6

◆

The Political Ecology of Logging and Related Activities

C utting rain forest trees is nothing new. The use of rain forest wood has been traditional for most human societies in contact with these ecosystems. But the European invasion of tropical lands accelerated the wood cutting enormously, as tropical woods began contributing to the development of the modern industrial society.[1]

The direct consequences of tropical rain forest logging, apart from the obvious and frequently spectacular visual effects, are largely unknown. Some facts are deducible from general ecological principles, and a handful of studies have actually measured a few of the consequences, but a detailed knowledge of the non-trivial direct consequences of logging is lacking.

What can be deduced from ecological principles is not that tropical forests are irreparably damaged by logging, but quite the contrary, tropical forests are potentially quite resilient. While this is perhaps debatable, most of the debate centers on how fast a forest will recover after a major disturbance such as logging (assuming nothing further occurs like subsequent fires, or agricultural activities), not on whether it will. The process of ecological succession inevitably begins after logging, and the proper question to ask then is, "How long it will take for the forest to recover."

In analyzing the effects of logging, we cannot assume a homogeneous process. There are a variety of logging techniques, some likely to lead to rapid forest recovery, others necessitating a longer period for recovery. For example, local residents frequently chop down trees for their own use as fence posts, charcoal making, or dugout canoes (figure 6.1). Forest recovery after such an intrusion can be thought of as virtually instantaneous, since the removal of a single tree is similar to the creation of a light gap, a perfectly natural process that happens regularly in all forests. At the other extreme is clear cutting, the extraction of all trees in an area. Though the physical nature of a clear-cut forest is spectacularly different from the mature forest, from other perspectives the damage is not quite so dramatic. The process of secondary succession, which begins immediately after such logging, leads rapidly to the establishment of secondary forest. A great deal of biological diversity is contained in a secondary forest, and indeed, a late secondary forest is likely to appear indistinguishable from an old-growth forest to any but the most sophisticated observer, even though it may have been initiated from a clear cut. It may even be the case that large expanses of secondary forest contain more biological diversity than similar expanses of old-growth forest.[2] No studies thus far have been able to follow such an area to the point where it returns to a "mature" forest again.[3] One could estimate something on the order of 40 to 80 years needed, perhaps, before the forest begins to regain the structure of an old-growth forest.

Probably the most common type of commercial logging is not the clear cutting described above, but rather selective logging. In an area of tropical forest that may contain 400 or more species of trees, only twenty or thirty will be of commercial importance.[4] Thus a logging company usually seeks out areas with particularly large concentrations of the valuable species and ignores the rest. Often the wood is so valuable that it makes economic sense to build a road to extract just a few trees.

Figure 6.1.   Informal uses of rain forest wood. Top photo is a dugout canoe being made from a downed *Ceiba pentandra* tree. Middle photo is of a completed dugout canoe. Bottom photo, a young Rama boy takes bags of charcoal to market in a dugout canoe. All photos from eastern Nicaragua.

There are three negative facts about selective logging that must be considered: road building, secondary damage, and deterioration of the stand. While direct damage to the forest from roads is considerable, it is the indirect consequences that have more important effects. Such roads offer access to the forest (figure 6.2). Hunters, miners, and peasant agriculturalists have new and extensive access to areas difficult to get to before. In most situations this aspect of selective logging is probably the most egregious for deforestation.

Figure 6.2. A recently logged area in southern Costa Rica. Note that there is forest on either side of the logging road. While many trunks had been removed, this particular operation left standing substantial amounts of vegetation, and the regrowth of the forest probably will be quite rapid. The road, on the other hand, will provide easy access for homesteading agriculturists.

Secondary damage from selective logging refers to the impact on non-target trees during the process of removing the targeted ones (figure 6.3). It is frequently extensive. Several studies have calculated that when a logging operation seeks only a small percentage of the trees in a forest, usually less than 10%, the incidental damage done by machinery moving logs through the forest, and by building roads, results in many of the trunks knocked down.[5] On the other hand we have observed small-scale operations with trunks hauled by oxen, in which the damage is significantly lower. It would seem obvious that the nature of the selective logging will partially determine the rate at which forests recuperate, which in turn is dependent on the extent of initial damage.

Figure 6.3. Bulldozer moves a recently felled log, scraping it along the forest floor, and in the process killing most if not all of the advanced regeneration of the forest, probably setting back the regeneration time of the forest significantly. Photo taken in eastern Nicaragua.

A final aspect of selective logging, deterioration of the stand, must also be acknowledged. Removing only the valuable species and genetic types while leaving behind the less valuable ones, results in an ever-decreasing fraction of trees in the recuperating stand that belong to the valuable species and genetic types. If a rotational system is initiated in which logs are cut every twenty or thirty years from the same site, it is easy to see how the stand will become less and less valuable as time goes by. In addition, if *all* the trees of a particular species are removed, that species can be driven to local extinction. This unfortunately occurs frequently with valuable species like mahogany. Furthermore, studies on the effects of selective logging on wildlife, which plays an important role in pollination and seed dispersal, have been contradictory and controversial.[6]

Reforestation and Tropical Plantations

Logging companies are frequently chastised because they are, "cutting more forest than they are planting," and it is increasingly common for conservation-minded people from the north to promote brigades of tree-planting volunteers to "restore" the rain forest. Such actions are sometimes misguided.

Rain forests contain countless species of plants and animals, and any attempt to restore them directly is absurd, even if all species were somehow available for restoration, which they usually are not. Furthermore, no one has anywhere near the requisite knowledge to advise how the restoration should proceed. How many tree species should be planted? Exactly where should they be planted? Should animals be introduced to disperse seeds? Should artificial perches be established to attract bird dispersers? Artificial bat roosts to attract bat dispersers? Should vines be planted or cut back? Given our current knowledge, calls for direct reforestation are at best naive.[7]

But the question of reforestation can be worse than simply naive. Contrast, for example, the "logged and left" area of the Danum Valley on the island of Borneo, in Sabah, Malaysia, with the "logged and reforested" area south of Kota Kinambalu, on the southwest side of Sabah. In the Danum valley there is no attempt to engage in any reforestation activities and both selective and clear-cut logging have evidently been extensive. Vast areas have been converted from old-growth rain forest to secondary succession. Much of the area appears devastated, but a closer look reveals an area devastated, but not dead. Pioneering tree species are beginning to form sections of diffuse canopy cover, and the process of ecological succession is well under way. The general area is currently a mosaic of various stages of forest regeneration, with old-growth forest near many rivers and streams, advanced second-growth forest in areas that were logged long ago, younger second growth in more recently logged areas, and scattered areas of very retarded vegetation associated with logging roads and skidding areas. A biological survey of the area would likely find a majority of the species that had been in the area before logging, but their numbers and phenological stage would be significantly different. Trees that may have been common before are now present only as seedlings and saplings; insects of the deep understory are probably now restricted to the patches of old growth near the rivers; and the larger mammals have probably gone elsewhere, or are hiding in those old-growth patches. For the most part, however, the forest is actively regenerating, and the species that were there before are still there, albeit in reduced numbers. In short, the area appears more devastated than it is.

In contrast, the forest concession managed by Sabah Forest Industries Inc., has a different management plan. The old-growth forest is being cut down and replaced with plantations of a few fast-growing species, particularly, *Acacia mangium*. The plantations form uniform stands of fast-growing trees which will be harvested on an eight-year cycle to feed a giant pulp and paper mill. Does this represent "reforestation"? This is not a simple question and involves some tricky cost and benefit considerations.[8] But from the point of view of restoring a natural rain forest, the Sabah Forest Industry program appears to be far worse than nothing at all. A forest of hundreds of species of trees, the myriad insects which feed on them, and the innumerable organisms that live in the immeasurably diverse litter, have been converted to a monocultural plantation. The loss of biological diversity has been almost complete. This is obviously by design, since the goal was to establish a plantation of Acacia trees.[9]

Frequently reforestation is precisely like the Sabah Forest Industries program. It is called "reforestation," but because the goal is to create a uniform product to feed a saw mill or a pulp mill, species composition is dramatically reduced and then maintained in this restricted state. This is a plantation, and from the point of view of biological diversity a plantation is more like a cattle pasture than a rain forest.[10] Furthermore, the fact that most of these fast-growing tree plantations are harvested in short (7–10 year) cycles, brings up some questions about sustainability. Can a plantation like this be sustained over the long run? What will happen to the soil after the trees are cut and before the newly planted trees begin forming a protective canopy layer?

This brings us to the notoriously difficult issue of tropical tree plantations. Unlike temperate forests, which, because of their low species diversity appear almost like plantations, tropical rain forests are dramatically more diverse. This diversity makes them both valuable and not valuable — valuable in that they contain scattered and rare high-value timber as well as many species of plants and animals, which may have worth in the future; not valuable because the vast majority of the trees presently have little commercial value. For this reason it is tempting, from an economic perspective, to turn the forest into one dominated by just the high-valued trees. Thus far this has been a difficult proposition the world over. Mahogany, for example, has been planted in monocultures and usually fails because of a species of insect that attacks the growing stems.[11] The insects apparently have difficulty finding the scattered trees that occur in a natural rain forest, but travel easily from tree to tree in a plantation.

On the other hand, if tropical tree plantations can be made successful, they would provide much more valuable wood per hectare than natural forests. A successful tropical plantation would take considerable pressure off natural forests. Accordingly, it is a mistake to classify tropical plantations as useless. However, plantations must not be confused with natural forests. Creating a plantation is not the same as reforestation. Sometimes, perhaps most of the time, the attempt to reforest rain forest lands is really an attempt to establish tree plantations. This can take on pernicious political overtones. Several years ago in Nicaragua we were told a Swedish company was planning to establish large plantations of *Eucalyptus* and pines in a former rain forest area. This was in response to criticism from their home country's rain forest preservation movement that they were "cutting down forest and not reforesting." The confusion between creating a plantation and reforesting a rain forest allowed the company to continue logging rain forests, while satisfying their political critics.

Another interesting example is the Ston Forestal company and its tree plantations in southern Costa Rica. Using the Australian Melina tree, Ston has established 20,000 hectares of tree plantations,[12] sometimes cutting secondary forest in order to do so. The plan is to harvest the trees for wood chips. Since the Costa Rican government offers tax credits for reforestation, the company insists that its efforts to set up plantations are actually "reforestation" efforts. Again, the confusion between reforestation and planting plantations has significant political consequences.

While there may be some exceptions, generally the best way to reforest a logged rain forest is to leave it alone. Rather than focusing attention on trying to actively restore a deforested area, energy would be better spent focusing on the nature of the logging operations themselves, and on what happens after logging, for example, secondary agricultural incursions, and forest fires.[13]

Disturbance and Recuperation in Rain Forests

A discussion of the political ecology of deforestation must occur within some basic ecological framework. As Blaikie and Brookfield[14] emphasize, notions like degradation (and by implication deforestation) must be carefully defined, for complicated subtleties in definitions suggested by different people and groups easily lead to confusion. To a botanist or ecologist, a deforested area may mean any forest that has suffered from logging in the past; while from the point of view of a forester, it may be an area with no harvestable timber. Or, deforestation may refer to areas dedicated by social custom to be devoid of trees (e.g. cotton fields, cattle pastures). Coupled with this difficulty of definition is the growing acceptance among ecologists that disturbance is not generally a bad thing, but rather part of the normal ecology of a region.

The role of periodic disturbance has long been recognized in ecology as an important organizing force.[15] Certain ecosystems, such as prairies, savannas, and many forests, derive their structure from periodic fires. Tree falls create light gaps in a forest, initiating a process of succession on a micro-scale, and possibly causing the high species diversity of tropical rain forests. Catastrophic wind damage is a well-known determinant

of certain structural features of forests. Our work in the Atlantic lowlands of Nicaragua suggested direct and relatively rapid regeneration of
the forest after a hurricane (figure 6.4).

Figure 6.4. Natural disturbance is a recurring event in many lowland rain
forests. Eastern Caribbean forest before disturbance (photo on top) and after
passage of Hurricane Joan (photo on bottom).

Whether the main form of disturbance is a hurricane or a natural tree fall, the implications of disturbance ecology are the same. It should be possible to design an extractive system that imitates natural disturbances in such a manner that recuperation of the forest is relatively rapid. Pickett[16] has suggested that the message of disturbance ecology for any sort of production activity might be to tune the production activities to mimic as much as possible normal disturbance events. The Palcazu project in Peru, where strips of forest are cut so as to mimic natural tree-fall gaps, represents an implementation of this idea.[17] Perhaps foresters can gain some insights by studying the sort of damage done by a hurricane and how the forest recovers from it.[18] For example, a hurricane may topple or truncate a vast majority of the large trees, but do little damage to the understory, including the saplings, which will eventually form the new tree layer. In a logging operation with heavy machinery, the understory is frequently damaged severely, sometimes virtually eliminated. The damage logging machinery does may be far more significant than the logging operations themselves.

To some extent the outlines for new programs can be seen in the philosophies espoused (though not necessarily practiced) by some foresters, rather than the one-dimensional programs commonly promoted by traditional conservationists. The recent summary of timber extraction in many tropical areas by Poore,[19] or the excellent history of forestry by Westoby,[20] are two of many possible primers. The point on which these and other forestry analyses agree, is on the role of secondary activity in the logging operation, most importantly the invasion of agriculture into logged areas.

However, the story is yet more complicated since agricultural conversion is not a homogeneous process. The nature of the agricultural activity is clearly an important variable. A small traditional peasant farm creates an environment that appears to share some features of tropical forest structure, with fruit trees forming an upper layer, other perennial crops forming a subcanopy layer, shade-tolerant plants in the understory, and patches here and there where the basic grains are grown in "light gaps."[21] When abandoned, such a farm is likely to revert to forest rapidly, provided there is a forest nearby to provide the necessary seeds. A peasant farmer who uses chemical pesticides may do further damage to the ecosystem, so that a return to forest after abandonment is somewhat more retarded. A small commercial farmer who uses machinery, chemical fertilizers, and pesticides will do yet more damage, and a banana plantation or cattle ranch is likely to set back the process even further.

These observations lead us to the generalization that deforestation often proceeds as a three-stage process. Timber is extracted for commercial purposes, a disturbance whose recovery time is on the order of up to 50 years, and whose direct damage to biological diversity is only slight. Subsequent to timber extraction, the area may be converted to peasant or traditional agricultural activities, representing a disturbance whose recovery time is on the order of 50 to 150 years, and whose direct damage to biological diversity is significantly higher than the original logging operation. Subsequent to either traditional agriculture or timber extraction may be the introduction of intensive modern agricultural activity, representing a disturbance whose recovery time is on the order of 100 to thousands of years, depending on the details of the disturbance. This general process is summarized in table 6.1.

Table 6.1. The approximate damage to biodiversity and estimated time to recovery of various forms of human activities, based on general comments in the literature and our own experience. All figures are very rough guesses, as indicated in the text.

Activity	Damage to biodiversity	Years to recovery
Informal wood harvesting	nil	0-5
Selective logging	nil to minor	10-20
Clear cutting	minor	20-100
Traditional agriculture	minor to significant	50-150
"Modern" agriculture	dramatic	200 - ?

1Westoby, 1989.

2Some ecologists think that the actual number of species in an ecosystem increases as ecological succession proceeds, but only to a point. After that critical point, the diversity actually decreases, leading to the conclusion that a very old forest may be less diverse than a younger one.

3There is a problem with the definition here. Most ecologists today eschew the notion of a "mature" forest, and simply speak of "old growth." The notion of maturity implies something about a directed developmental sequence that does not fit well with what we now know about tropical rain forest succession.

4There are exceptions to this rule. Many swampy forests are characterized by the presence of only a few species. The biggest exception are the Southeast Asian Dipterocarp forests, where the vast majority of trees in the forest belong to a single plant family, characterized by very large and straight trunks, a logger's delight.

5Johns, 1985.

6ibid; Johns, 1991; Karr, 1971; Michael and Thornburgh, 1971; Webb et al, 1977.

7On the other hand, there is much research that is currently underway concerning this topic (e.g. Butterfield, 1990; Espinoza and Butterfield, 1989).

8The anticipated benefit of such plantations is that they will provide much greater wood yields, thus taking pressure for harvesting off of the natural forests.

9More recently the practice has apparently been halted, and Acacia plantations may be a thing of the past in this area (J. Anderson, personal communication).

10Lohmann, 1990; Lugo, 1992.

11The organism is called the meliacian stem borer (*Hypsipila sp.*) and attacks any plant species in the family Meliaceae. Its effect is to turn the tree into a small bush, since the insect bores into the growing tip, causing the tree to sprout several other growing tips at the base of the damaged one. This process continues until you have a mahogany bush instead of a mahogany tree.

12Interview with representative of Ston Forestal published in the Costa Rican Magazine *Panorama del Sur*, May 1993, pg 13.

13There are, on the other hand, several studies under way attempting to accelerate the process of regeneration of rain forests after logging (e.g. Huss and Sutisna, 1991; Maury-Lechon, 1993).

14Blaikie and Brookfield, 1987.

15Mooney and Gordon, 1983; Pickett and White, 12985; Sousa, 1984. This point was covered with regard to tropical rain forests in Chapter 2.

16Picket and White, 1985; and personal communication.

17The Palcazu project was designed according to the theoretical constructs of forest ecologist Hartshorn. Strips of forest were harvested in a clear-cut form, attemping to mimic the conditions normally found in forest light gaps (Hartshorn, 1987).

[18]Perfecto et al., 1994; Boucher et al., 1994; Yih et al., 1991.
[19]Poore, 1989.
[20]Westoby, 1989.
[21]See Chapter 7.

7

◆

Rain Forest Conservation: The Direct or Indirect Approach?

I n 1990 when World Resources reported on deforestation rates for most countries in the world, some new and surprising data became available to environmentalists. In its rush to free its terrain of trees, Costa Rica claimed the dubious prize for First Place. For almost a decade Costa Rica, partly because of its proximity to revolutionary Nicaragua, had received the bountiful largesse of the Reagan administration. Development aid for Costa Rica was enormous, part of the same overall strategy that brought war to Nicaragua.[1] Concomitant with this influx of U.S. dollars was a conservation movement, strongly influenced by North Americans, and with few parallels in the Third World. Combine this conservation movement with an enormous amount of money and one would expect spectacular results. Indeed Costa Rica's conservation policies suggest exactly that: approximately 27% of its national territory lies under some sort of protected status; it has arguably the most impressive national park system in the Third World (12% of its total land area[2]); and boasts some of the most progressive forestry laws in the world.[3]

Conservation policies in Costa Rica were based on the assumption that rain forest destruction could be curbed by 1) buying up and protecting

large tracts of land; 2) passing legislation; 3) securing large sums of project money from foreigners; and 4) mounting a massive public relations
campaign. Alas, all this was done with little concern about the impact of
some crucial socioeconomic factors, such as one of the most uneven land
distribution patterns in Central America.[4] The cherished belief was that
these policies were working. But the World Resources figures reported in
the summer of 1990 showed otherwise. Despite Costa Rica's policies, during the 1980's their forests came down at an alarming rate of 7.6% per
annum.[5] Such destruction certainly represents a reality check for all rain
forest preservation advocates. Not only was deforestation higher than
anyone had anticipated, it was the highest in the world!

   Preserving the world's rain forest has become *de rigueur* in the developed world. While it has been difficult to ignore such readily available
facts as the militant socialist agenda advocated by Chico Méndez,[6] the
western rain forest conservation movement has nevertheless remained
largely a bourgeois business. While we should perhaps have been asking
questions about the Indonesian military government and its transmi-
gration program to Kalimantan, or the Guatemala military's reported
use of herbicides to clear vegetation in areas where guerrillas were suspected to be hiding, or Nicaragua's progressive land reform program,
Costa Rica, for the insiders in the rain forest preservation movement,
represented a far cleaner story. In Costa Rica, no military intervention,
no socialism, no revolution, no anti-U.S. sentiment clouded the picture.
Bird watching and nature appreciation tours were touted in the travel
sections of all major U.S. papers. In addition, Costa Rican authorities
were amenable to advocating all the goals and programs of the mainstream rain forest conservation movement. Consequently the past
twenty years have seen the truly remarkable growth of a conservation
program, but one whose aims and advocacies have remained almost
totally isolated from some major socio-political questions plaguing
Costa Rica. And this isolation is precisely the problem.

   There has been, we believe, a general failure to address the underlying
causes of rain forest destruction. Our purpose — hopefully obvious by
now — in writing this book, is to note and acknowledge those root —

causes. We are well aware that our analysis is at odds with much of the international community that seeks to preserve rain forests. We do not despair at this incongruity. The problem of rain forest destruction is far too important to leave to misguided proposals and ineffective solutions.[7] It is an unfortunate truth that "empire building," careerism, and even economic self-interest, sometimes drive conservation programs, and foster "analyses" which systematically exclude a search for the real root causes. Perhaps this is because these root causes create conditions in other spheres of life, which the conservationists would not like to see challenged. Perhaps the same political arrangements, which provide conservationists with the privilege to ponder such weighty questions as, for example, biodiversity, also create the impoverishment that forces peasants to cut down rain forests. If so, it would be prudent to keep those political arrangements out of the spotlight — for the mainstream conservationists, that is. But we, as the Lorax,[8] seek to speak for the forest. For that reason we intend to focus the spotlight on exactly those underlying causes.

In this chapter we draw attention to these political issues in the hopes of challenging what seems to have become a myopic and elitist view of rain forest destruction — a view often, unfortunately, tailored more to careers in conservation biology and the maintenance of good relations with potential funders, than to a sincere desire to stop the progressive loss of the world's rain forests. We first revisit the *Sarapiquí* region of Costa Rica as described in Chapter 1, and detail some of the complexities that exist there. We then compare that site with a similar one in Nicaragua, elaborating patterns that existed there mainly during the 1980's. Our intent is to demonstrate first, that the main stream rain forest preservationist view of the issue is hopelessly superficial, and second, that an analysis of the entire matrix of socio-political-ecological forces is essential to understanding what in fact causes, in these cases, the destruction of rain forests.

The *Sarapiquí* Revisited

This book began with the example of the banana expansion in the *Sarapiquí* region of Costa Rica. It is fitting now that we return to that example (refer to the map in figure 1.2). As anticipated, the situation is deteriorating rapidly. The current players are an international tourism company, a well-connected conservation research organization, two local political groups, several absentee landlords, and some fruit companies (figure 7.1).

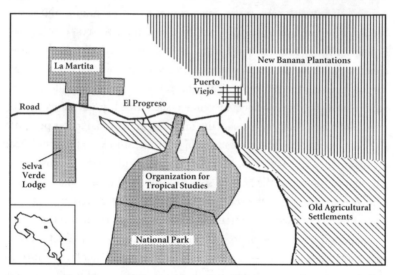

Figure 7.1. Sketch map of the region surrounding the town of Puerto Viejo in the *Sarapiquí* county of northern Costa Rica (insert shows location of area of detail), showing approximate locations of sites mentioned in text.

The first local political group is a large, highly organized group of homesteaders in a community they call *El Progreso*. Their story began some fifteen years ago when a North American who owned a very large ranch in the area, was indicted on drug trafficking charges.[9] He fled the country. According to Costa Rican law, an absentee landlord can have his land expropriated if Costa Rican nationals establish a homestead thereon, which is exactly what happened. A Costa Rican couple who had worked for the North American, established themselves as homesteaders and petitioned the appropriate government agency for title to the land. Meanwhile the North American got wind of what was happening and solicited the aid of a lawyer friend in Costa Rica. The lawyer apparently filed papers aimed at incorporating the ranch into a group of farms he already owned. At the same time the Costa Rican couple renamed the farm *Gerika*.

Thus the farm, now known as *Gerika*, came to have two or three owners: the indicted North American who left the country, his lawyer friend, and the Costa Rican couple. Then in August of 1993 tragedy struck the Costa Rican couple. The wife murdered the husband and then killed herself. In such circumstances Costa Rican law and tradition says that if there is a known long-term caretaker of the farm, ownership passes on to him or her. In this case a woman had been the caretaker, and she quickly laid claim to the farm. But there is another Costa Rican law that allows for homesteaders to occupy unused land. With the bizarre death of the Costa Rican couple, whom most locals had assumed were the owners, homesteaders saw the opportunity to seize what had been for quite some time unused land. Only weeks after the couple had passed away, homesteaders invaded part of *Gerika*, calling their new community *El Progreso*.

The invasion was not without violence, and the combination of angry homesteaders, private guards, and Costa Rican security forces produced some ugly incidents, including one death.[10] But a year after the establishment of the homesteading site, *El Progreso* was home to 350 families who had obtained legal "right to possession," the first step towards getting official title to the land under the Costa Rican homesteading act.

Several interesting facts highlight the *El Progreso* affair. First, some unknown number of the homesteading families are Nicaraguan (local estimates range from "not many" to "almost all"), most having been attracted to the area by the promise of employment in local banana companies. Second, the back border of the farmable land claimed by the homesteaders abuts directly on property claimed by the Organization for Tropical Studies, one of the major conservation players in the area. Third, many in *El Progreso* are currently employed by the banana companies, which send busses to the community each morning to pick them up, and many others are ex-employees of the banana companies. Finally, evidently the real estate they have occupied is fairly rotten. While the members of the community are optimistic about their dreamed-of farms, we doubt much will come of their agricultural activities. The soils seem to be old alluvium, but are probably very acid and lacking in nutrients.

The second major local political group is the Sarapiquí Association for Forests and Wildlife. Their goal is to "promote the conservation and sustainable development of the natural and human resources in the county of *Sarapiquí*."[11] The association was formed by local residents with the help of a non-governmental organization that provides legal assistance to local grass-roots groups seeking to develop conservation projects.[12] The Sarapiquí Association's immediate objective is to purchase a piece of land called *La Martita*, whose Colombian owner has applied for a logging permit from the government. As the association points out, several biological reserves and stands of old-growth forest exist in the *Sarapiquí*, but all are owned by foreigners. With the development of the ecotourism industry, local residents have garnered some benefit as employees of ecotourism entrepreneurs, but there seems to be no example of a local resident who is owner of a successful ecotourism establishment. The goal of the Sarapiquí Association is to use *La Martita* as a community biological preserve where locals can begin developing their own ecotourism operations (see figure 7.1).

Much of the *Sarapiquí* is still owned by absentee landlords, both Costa Rican and foreign. Excluding the banana companies, the main activity on these lands is cattle ranching, although many still contain substantial tracts of rain forest. Cattle produce very poorly in the region, and it has been suggested that the only reason they are kept on the land is to claim that the land is in "use" so it is not subject to invasion by homesteaders. *La Martita* is one such operation. The part of the property that connects with the main road is in cattle pasture and boasts a sign in front that reads "Warning, vicious dogs." The back of the property, the forested part, is not accessible from the main road, so it has not yet been a target of homesteaders. That is likely to change if the current owner sells the timber on the land and a logging company constructs logging roads into the area. It is not out of the question to envision another group of ex-banana workers establishing a homesteading community here, if the opportunity arises. If the association is successful in purchasing the land and preserving the forest, such an invasion is far less likely.

The international tourism company is Holbrook Travel Inc., and they own the Selva Verde Lodge (see figure 7.1). Styled after the highly successful nature tourism industry of Kenya, this lodge has up-scale facilities that shield the tourists from the local people, except for those who work as cooks, maids, or maintenance workers. This impressive facility includes the largest privately-owned nature reserve in the area, and is one of the largest employers in the local ecotourism business. Local people most frequently cite the Selva Verde Lodge when they speak about foreign ownership of the local ecotourism business.

The other large ecotourism operation is the Organization for Tropical Studies, OTS, a consortium of universities in the U.S. and Latin America, with some fifty members. Its goals are research and education in tropical biology, and it is one of the most famous organizations of its kind in the world. Because OTS attracts rich foreign ecotourists and is owned mainly by non-Costa Ricans, the organization has come to be viewed by many local people as similar to Selva Verde. Such a perception is natural.

Finally, and still the predominant feature of the area, are the banana companies, dominated by the multinationals Chiquita, Dole, and Geest. As we predicted in Chapter 1, the expansive phase of their operation has ceased, and many workers have been laid off. These workers and their families are now living in homesteading communities like *El Progreso*, homesteading individually in various corners of the landscape, seeking employment wherever they can get it (not much in the area other than the banana plantations), and increasingly migrating to San Jose, the capital. Most recently the situation of the workers has gotten so bad that the all but defunct banana unions have reappeared on the scene, and in May 1994 a short strike against the Geest plantations shocked the country.

This then is the current landscape. Ecotourism and international conservation groups rapidly losing credibility among the local population; banana companies attracting workers to the region and dumping

them onto the landscape; absentee ranchers hiring armed thugs to protect their land from homesteaders; international tourists who can spot
the rarest of birds through their binoculars yet cannot see the desperate
and hungry people just across the road; 350 hopeful and optimistic families in a new community with a utopian plan to turn acid soils into productive agriculture; and a weak, largely-ignored local conservation
group trying to get something for Costa Ricans out of ecotourism.

The current landscape would seem to leave little reason for opti-
mism. The dilemma, we believe, is that analysis of the problem has been
made in a fragmented fashion by the various interest groups. Local conservationists are concerned with preserving whatever forest remains and
see the homesteaders as conniving ne'er-do-wells. Some conservationists even attribute rain forest destruction to overpopulation. Foreign
ecotourism companies pay lip service to conservation, but cease even
that when their profits are threatened. Homesteaders see large tracts of
land unused by foreigner owners and view their own landless situation
as unfair. Gaining property title to a piece of logged-over forest seems
logical and fair to them. The central government continues its policy of
promoting the expansion of bananas, even while promoting conservation programs in other areas of the country. And lastly, the banana companies see the conservationists as romantic idealists who simply do not
understand the nature of business, which is to make profits; nor the
need of the country, which is to generate revenues to pay off their international debt.

If, however, the problem is viewed in all of its complexity, and a plan
devised that takes all parties into consideration, there is much that can
be done. We envision, tentatively, a three-part program. First, local people need jobs. We regard the question of food security as the underlying
basis of any sound conservation policy. To this end, the labor practices
of the banana companies must be regulated. No longer should they be
allowed to hire workers for eighty-nine days and then fire them, because
the companies become liable for various social security benefits after
ninety days. Working conditions must be improved, and above all job
security must be instituted. While there is little that can be done about

1) job security - banana
2) agriculture

3) employment in ecotourism

agriculture

the banana plantations already in existence, a moratorium should be immediately announced concerning the further expansion of such production technology.

Second, there are clearly not enough jobs in the banana industry to employ everyone in the region. It should be noted that during the banana expansion of the early 1990's, the banana companies asserted there were not enough workers in the region, and in coordination with the Costa Rican government, brought in Nicaraguan workers. Rumors of available jobs attracted people from all over Costa Rica and Nicaragua, dramatically increasing the region's population. The only obvious beneficiary of this chaotic migration process has been the banana companies, which now enjoy a surplus of workers.

The reality remains that not everyone will get a job in the banana industry. Existing homesteader communities, beginning with *El Progreso*, need technical aid and bank credit to make their farming operations as viable as possible. It should be noted that viability may not, in fact, be technically feasible. Despite twenty-five years of active biological inquiry in the area, only a tiny fraction of that research is even theoretically relevant to finding solutions to the many problems faced by farmers in the region. Indeed, if the current agricultural areas of *El Progreso* turn out to be difficult to farm on a continuous basis, *El Progreso*'s farmers may expand their agricultural activities into the forested areas that border their farms. This includes some of the forested land claimed by the National Park, the Organization for Tropical Studies, and the Selva Verde Lodge. It would seem that making *El Progreso*'s agriculture successful should be an obvious priority, even for the ecotourism ventures.

Third, local people must directly benefit from the increase in ecotourism in the region, not just as janitors, cooks, and badly paid tour guides, but as entrepreneurs and owners of lodges and reserves. This probably means support for the Sarapiquí Association for Forests and Wildlife, especially with their current efforts to purchase *La Martita* and turn it into a community-owned forest reserve. Part of this general goal can be easily coupled with the agricultural imperatives for the region. Ecotourists are often interested in seeing alternatives to destructive agriculture and the people of *El Progreso* already have an impressive conservation ethic. Several homesteaders independently offered the opinion that deforestation was the country's biggest problem right now, and that ecotourism should be promoted. They see potential jobs for themselves in ecotourism. And, if they begin an experimental program of sustainable agriculture/agroforestry, their land might produce more than just corn and beans, it might produce tourist dollars.

What chance does this three-part program have for success? Given the constraints of the real world, probably very close to zero. But each component is worth pursuing. The alternative, foreign conservation organizations and ecotourism companies dominating all the forested land, banana companies dominating the remainder and hiring and firing workers chaotically, will certainly not work in the long run. Poor people need to feed themselves and their families. They need steady jobs or access to good agricultural land and technology. If they have no job and no hope of getting one, if they are excluded from the promise of the bright future they can see the foreigners already have, if their crops continue to fail on the small piece of rotten land they have grudgingly been allocated, they will do whatever they deem necessary to survive. Even if that means cutting and burning primary forest on preserved land.

By Contrast, the RAAS of Nicaragua[13]

The story in Nicaragua is quite different from Costa Rica, although it also illustrates features of the general pattern we have described. The Atlantic coast of Nicaragua contains the largest remaining tract of lowland rain forest in Central America,[14] with significant closed-canopy forest, or at least patches of it, extending from the border of Honduras all the way south to the border of Costa Rica (see map in figure 1.2). There are, however, several foci of agricultural expansion. For the past few decades the largest of these agricultural frontiers has diffused eastward toward the area surrounding Bluefields on the Atlantic coast. Since 1990, the rate of expansion has increased dramatically, perhaps aided by the devastating effects of the hurricane of 1988 and the elections of 1989. The agricultural frontier has effectively cut the Atlantic coast rain forest into a northern and southern section, and is now expanding northward and southward (see figure 1.2). If this pattern continues we can expect the destruction of all but small islands of protected forest, with little potential for stable economic activity for the local population.

The majority of the upland area surrounding Bluefields (about 400,000 hectares) was originally covered with lowland tropical rain forest, almost all of which was severely damaged in 1988 by Hurricane Joan. The forest is now in various stages of ecological succession. Despite the severe destruction, our studies indicate that natural resprouting and direct regeneration have resulted in a habitat well on the way to mending itself. Small peasant farms dot the area and continue expansion into formerly forested land. A large sugar plantation is located north of Bluefields, and an old extensive cattle operation covers an upland area known as *Loma de Mico*. For the past decade the sugar operation has rarely been profitable, and the cattle operation was virtually abandoned in the mid 1980's. Finally, a large African oil-palm plantation was planted in the mid 1980's and is producing at approximately 25% productive capacity.

The human cultural history of the Atlantic coastal area is extremely complex. The bulk of the population is about evenly divided between African Americans brought from Jamaica as slave laborers in the 1800's, and Mestizos who migrated from the Pacific coast during the past century. Additionally, three small Native American groups inhabit the area: the Rama; the Sumos who have small communities north of Bluefields; and the Miskitos, most of whom live in the North Atlantic zone, with small numbers in the South Atlantic zone, living in either small villages or as individuals among the other cultural groups. Another ethnic group, the Garifuna, originally derives from escaped African slaves. While all groups maintain strong cultural identities, they are also united in a coast culture and regard themselves as *costeños* (coast people).

The political panorama is as complex as the cultural. The well-known friction between the Sandinista party and the UNO (United Nicaraguan Opposition) is complicated by the presence of YATAMA (another political party representing mainly the Miskitos), and the entire political framework is permeated with the question of autonomy. Historically isolated from the Nicaraguan mainstream, the Atlantic coastal region has always been fiercely independent in politics and economics. In 1989 the first autonomous administration was elected, and — at least on paper — the RAAS (South Autonomous Atlantic Region) is supposed to be independent of the central government administration. This formal independence extends to decisions on all natural resources, including rain forests.

The RAAS encompasses the southern half of the Atlantic coast of Nicaragua (see map in figure 1.2). Economically, it is something of a disaster. While Nicaragua is second only to Haiti in poverty in this hemisphere,[15] the Atlantic Coast is the most underdeveloped region of Nicaragua. Local government officials claim an unemployment rate of over 90% in Bluefields, and a short visit to the area convinces even a casual observer that this is indeed the poorest region of Central America.

During the Somoza years (1930–1979) the area was relatively isolated, and while Nicaragua as a whole has been reported to have experienced massive deforestation during that time, little evidence exists for such deforestation in the RAAS in particular. During those decades, the agricultural frontier slowly expanded towards Bluefields, the large sugar plantation and cattle operation opened up, and timber extraction, primarily by North American companies, was more-or-less continuous. Though it is not known for certain, it is probable that, as elaborated above, the local logging operations were followed by peasant agriculture.

Following the Sandinista Revolution of 1979, two major forces relevant to tropical forests came into being. First, the contra war waged by the Reagan administration made it especially difficult to engage in large forestry operations. The contra war also made it difficult, in areas constantly threatened by fighting, for peasants to homestead or create new agricultural plots. However, casual interviews with people in Bluefields do not suggest that large numbers of landless peasants were champing at the bit waiting for the war to end so they could carve out a homestead from the remaining forest. To the contrary, the landless peasantry seems to have largely disappeared due to the second relevant force: the Sandinista administration's agrarian reform program. The rate of deforestation in the region was relatively low during the eighties.[16]

The elections of 1989 changed everything. The progressive agrarian reform of the Sandinistas was reversed, and once again landless peasants were forced to seek new homesteads, frequently after having been forcibly removed from land acquired under the Sandinista reforms. As part of a deal with the new government, so-called "poles of development" formed new bases of agricultural expansion. Former contra rebels were given forested land to distribute among their "soldiers" and their families to establish agricultural communities. A typical pattern consists of two or three years of cultivation after cutting a piece of forest, followed by a ten-year fallow period, followed by a single year of cultivation, and a strong desire to move on in hopes of a better piece of real estate under the next parcel of forest.

The relative stability the Sandinista agrarian reforms provided now seems to have been destroyed. We have been traveling in this area since 1982, and the changes since 1989 have been frightening. Before the 1989 elections there was a great deal of debate on questions like: use of the natural resources of the area; the role of the state in regulating chain saws; how the state logging company (the only logging company operating) should be responsible to local communities; and how agrarian reform must extend into the technical sphere to provide peasants an alternative to burning for land preparation. By 1993, the constant drone of chain saws could be heard anywhere there was forest. Three lumber companies were actively engaged in cutting trees (openly flaunting a ban on cutting), and central government figures were accepting bribes for allowing logging concessions.[17] Fires were everywhere as landless peasants burned yet another piece of forest, and former contra bands, apparently frustrated with their inability to produce anything on the poor soils of the region, engaged in robbing passing boats and fighting with one another.

In addition, the economic situation has deteriorated significantly. The peasantry, which in the past could hope that the government would help in times of crisis (e.g. if the crops failed there was reasonable expectation that the government would supply the community with basic grains), has become despondent and is often forced to do anything possible to eke out a living, usually with little success. A typical story was related to us by the residents of a peasant community on the Patchy River, north of Bluefields. The community, a group of ex-contra rebels, had been resettled in the area under the leadership of their former commander, whose *nom de guerre* was Ranger. Ranger contracted with outside contractors to purchase the mahogany and rosewood from the forest behind their small settlement. The community had neither beasts of burden, nor tractors, so Ranger paid the peasants the equivalent of $3.50 per trunk (which would retail for at least $1000 on the world market), to haul them down to the river dock by hand. In the end, to add insult to injury, Ranger disappeared with the logs and didn't even pay his men the $3.50 per trunk he had promised.

The Lessons of the Costa Rica/Nicaragua Comparison

That the rain forests of Central America are being destroyed is a point on which there is no debate. The causes of their destruction are considerably more debatable. Many biologically-oriented analysts tend to tie the destruction to the simple metaphor of the rain forest as a commons, and overpopulation the driving force. This might lead to the (erroneous) conclusion, for example, that the development of effective birth-control programs in the *Sarapiquí* would stop small farmers from seeking farms in the area. The reality is different. Since workers are actively recruited into the area by the banana companies, condoms are unlikely to solve the problem of rain forest destruction. Indeed, as far as the banana producers are concerned the area remains underpopulated.

The alternative explanation, more complicated and nuanced than overpopulation, has been presented above and in Chapter 1. It takes its framework from a variety of classic ideas, and sees rain forest destruction as the consequence of the interaction among several forces operating within a particular world system. Three modes of production (forestry, peasant agriculture, and modern agriculture), interact with

explanations for destruction   *Cultural*
                                                            Cultural

two socioeconomic groups (export agriculture bourgeoisie, and rural
peasant or proletariat) in the context of a national and international
political structure in an ecological matrix. It is impossible to understand
deforestation if one remains at only one or another of these levels, since
the cause is located in the overall structure.

A comparison between Nicaragua and Costa Rica during the early
1980's is illuminating. Both countries had significant areas of rain forest
remaining, but Costa Rica lost rain forest areas at an extremely rapid
rate during the 1980's, while the rate of deforestation in Nicaragua was
much lower.

Nicaragua was distinct from the other Central American countries
during the 1980's. For one thing, the contra war mitigated against significant lumber operations, hence extensive access roads were not built. *war*
The war, while sometimes intense in places along the northern border,
was a more low-intensity conflict where the rain forests were concen-
trated on the Atlantic coast. Small bands of counterrevolutionaries were
scattered throughout the Atlantic lowlands, especially in heavily-forested
areas. Thus the region was not likely to be subjected to intense commercial logging, or even small peasant clearings during the war.

However, there was an even more important factor. The massive
agrarian reform program initiated after the 1979 revolution all but eliminated land-hungry peasants in Nicaragua. In 1978 36% of the land was *agr. reform*
in farms larger than 850 acres. By 1985 that fraction had dwindled to
11%. In 1978 there were no production cooperatives. By 1985, 9% of the
land and 50,000 families were integrated into the cooperative sector. By
1985, over 127,000 families had received title to their own land.[18] This
was an agrarian reform without precedent in the history of Latin
America.

Costa Rica's agrarian reform program was entirely different, bearing
the recognizable fingerprint of similar programs in El Salvador, and —
years before that — in Vietnam. Organized by the state agency IDA
(Instituto de Desarrollo Agrario), landless peasants were located on or
near undeveloped areas, only occasionally on the lands of large
landowners. Expropriation of large inefficient enterprises was totally

absent from the system, and land titles were given with twenty- or thirty-year mortgages, forcing farmers into cash-crop production. Traditional farmers who sought a plot of land, if they got anything at all, acquired a piece of marginal, undeveloped land and, for the first time in their lives, a bank debt.

Interviews with small farmers in Nicaragua and Costa Rica reflected the basic differences in agrarian reform programs. Costa Ricans emphasized the land tenure issue, voiced disquietude over their lack of land title, or their inability to pay recently acquired mortgages (many were quite surprised at their new debt), and focussed their economic concerns on the attainment of land security. Nicaraguans were the opposite, at least before the 1989 elections. While they had many legitimate gripes, lack of land was not one of them, and rarely did a Nicaraguan small farmer talk about needing a piece of land to call his own. The consequences with regard to pressure on rain forests are obvious.

With this model of land tenure in Nicaragua, one could have predicted that the pattern, so common in Costa Rica, of land-hungry peasants following logging roads into the forest, would not be a problem in Nicaragua. Recent electoral changes in Nicaragua have put this model to the test, since the new government made the rollback of much of the agrarian reform program part of its electoral platform. As expected, and as detailed above, the new class of landless peasants is busy following new logging roads, and clearing forests in southeastern Nicaragua, conforming quite well with our overall interpretation.

One point of comparison outside of Central America is worth making. The Caribbean nations of Puerto Rico and Cuba are the only tropical countries in the world to have experienced an increase in the area covered by forest in the past few decades,[19] though for different reasons. Because of the development program imposed by the colonial administration in the 1940's, Puerto Rico can hardly be categorized as an agrarian economy; and in view of its peculiar articulation with the United States (not of its own accord), Puerto Rico has become an industrial colony. In a grotesque parody of real development, Puerto Ricans have been converted into classical proletarians by massive federal money

Puerto Rico &
Cuba

transfers. Consequently, to speak of a rural peasantry in Puerto Rico is anachronistic, and any notion of a movement of landless peasants into forested lands is ludicrous. Industrial development — however artificial and imposed from afar without the democratic participation of the Puerto Rican people themselves — has transformed the class structure of Puerto Rican society to such an extent that the basic structure, which we assert causes deforestation, simply no longer exists there.

Cuba's development has been along a different line than Puerto Rico's, emphasizing socialist goals. The successes and failures of this line of development are obviously hot topics for debate, but a single feature of Cuban developmental strategy is relevant to the present discussion. Cuban peasants are not land-hungry. Because of the basic philosophical commitment to a secure economic environment for all members of society, the Cuban peasantry has either been absorbed into the urban sector, or has received land security on state farms, cooperatives, or private farms. Cuba simply has no landless peasants. It is worth noting the obvious at this point. The probability that either the Cuban socialist solution, or the Puerto Rican federal-subsidy solution will occur in Nicaragua, Costa Rica, or in any other tropical country, is vanishingly small. So what might we realistically expect in Central America? Indeed prospects for the future do not seem sanguine. If general world trends continue, and especially if the relationship between the developed and underdeveloped worlds evolves along the lines anticipated by the so-called "New World Order," we can only expect more of the same. The class conflicts that raged in Guatemala, El Salvador, and Nicaragua during the past twenty years are not likely to subside, although their specific form will change. With the threat of U.S. military adventure constantly on the horizon, and no longer balanced by the Soviet Union, we can expect continual increase in the power of bourgeois elements, and consequent further erosion of the land base of the peasantry.

Everything seems to be in place for the continued growth of landlessness and poverty. If industrialization cannot absorb the expanding landless peasantry, then a replay of past deforestation patterns seems unavoidable. And with the current international debt situation, and

continuing political conflict, the prospects for significant industrialization do not seem very bright. The vision of agrarian reform programs
like Nicaragua had in the 1980's has all but disappeared in its country of
birth, and is hardly a serious proposition in the other countries of the
region. Therefore, the basic rural program that would help stem the tide
of deforestation, agrarian reform, is not realistically on the horizon.
Industrialization, the only other option, seems just as far off.

This puts conservationists in a holding pattern dividing along the
same lines we introduced in Chapter 1. Mainstream environmentalists
have been concerned with accumulating large sums of money to purchase and protect islands of pristine rain forest, with little concern for
what happens to either the ecosystem or the human societies located
between those islands. As we indicated previously, we do not imagine
this strategy has much chance of working in the long run. The landscape
will be converted into isolated islands of tropical rain forest, drowning
in a sea of pesticide-drenched modern agriculture, with masses of landless peasants looking for some way to support their families. Unless the
pace of industrialization increases dramatically (something few
observers expect), these peasants are unlikely to be absorbed into the
industrial work force. Instead, they are very likely to begin homesteading
in those islands of protected forests.

The alternative strategy emphasizes the land *between* the islands of
protected forest, and acknowledges the interconnections in this complicated system. This sustainable development point of view conceives the
ecological side of the dilemma as a landscape problem with forests,
forestry, agroforestry, and agriculture as interrelated land-use systems,
and seeks to develop those land-use systems to maintain conditions of
production. The idea has seen much recent analysis,[20] and is perhaps
our best hope.

Past historical contingencies and current economic realities place the
Third World in a disadvantaged position for economic development
under the world system. Development specialists from all spheres of
political and economic influence have attempted to deal with the Third
World problem ever since it was recognized, and "development," irre-

spective of its sustainability, has not been common. The imposition of restrictions like sustainability only heighten that disadvantage. Quite the contrary, an advantage will accrue to those regions and countries that do not impose constraints on their development. Sustainability is just such a constraint.

So we end this chapter, first with a paradox and finally, with a note of pessimism. Stemming the tide of rain forest destruction requires not only development, but development that is sustainable. Obviously sustainability is not sufficient, but is a necessary requirement. In the New World Order rational planning is not anticipated at either national or international levels, while development will proceed fastest for those able to ignore constraints others either cannot or will not ignore. Sustainability is precisely such a constraint. Thus, although sustainability is required to save the rain forests, it seems unacceptable as a constraint on development, at least under conditions of the current world order.

Our note of pessimism, if the above paradox is not sufficiently pessimistic, is that the structures to which we lay blame for the destruction of rain forests are still in place, probably more solidly than ten years ago. The same modes of production (forestry, peasant agriculture, modern agriculture) interact with the same socioeconomic groups (export bourgeoisie, peasantry) in the same national and international political arena within the same ecological matrix. Forces in the national and international political arena have constructed an ideology in which the rain forest is assumed to be an externality that fits into and fortifies the overall political structure, thus creating an inexorable dynamic in which deforestation is an inevitable consequence. As long as present political arrangements survive, a solution simply does not seem possible. To the extent that conservationists and the conservation movement are part and parcel of those political arrangements, they remain part of the problem.

[1]The history of the 1980's in Central America has been covered in many volumes. For example, Rosset and Vandermeer, 1986; Barry, 1987; and Edelmen and Kenen, 1989, are excellent sources.
[2]Boza, 1993; Tangley, 1990.
[3]For a critique of these forestry laws see Thrupp, 1990.

4Seligson, 1980.

5World Resources Institute, 1990.

6Chico Méndez was the militant organizer of the rubber tappers in Brazil. His death at the hands of thugs hired by the cattle ranchers has galvanized much of the rain forest preservation movement. In addition to his well-known activities on behalf of the rubber tappers, Méndez was a leader in Brazil's Workers Party, and an advocate of socialism as the only ultimate solution to the problems of environmental deterioration and social injustice (Méndez, 1989).

7A similar point of view, associated with African wildlife, was presented by Bonner (1993).

8Children's author Dr. Seuss' mythical creature who spoke out against the ruthless exploitation of the trufula trees.

9This and the rest of the story of *El Progreso* comes from interviews with several residents of the community conducted in the summer of 1994.

10 *Tico Times*, July 9, 1993.

11Document prepared by the Sarpiquí Association for Forests and Wildlife (undated).

12There is some debate as to how strong the influence of foreigners is in the association (personal interviews with local residents, 1994).

13Most of this section is taken from Vandermeer, 1991, and Perfecto et al., 1994.

14In 1990 the FAO reported the following figures for rain forest extent in Central America (in 1,000 hectares): Costa Rica - 625; El Salvador - 33; Guatemala - 2,542; Honduras - 1,286; Nicaragua - 3,712; Panama- 1,802 (World Resources Institute, 1994).

15In 1986, Haiti's per capita GNP stood at $330, while that of Nicaragua was $790. For comparison, the Dominican Republic was $710 and Honduras was $740. All other Central American and Caribbean countries had GNP's larger than Nicaragua. By 1991, Nicaragua's GNP stood at $283, the worst in the hemisphere (Haiti stood at $375 at that point). We do not have more recent figures in Haiti, but subsequent to the ouster of Aristide and the economic blockade, we strongly suspect Haiti has returned to its unfortunate position as the poorest country in the hemisphere. Nicaragua is a close second.

16Vandermeer et al., 1991.

17This was a widespread claim in both the Atlantic coast and Managua. Obviously its strict verification is hardly possible under current political circumstances.

18Kaimowitz, 1986.

19Current estimates suggest that about 18% of Cuba is covered in forest, which clearly represents an increase from the figures of 1959 (Rosset and Benjamin, 1994). Some of that area is clearly under plantations of trees and thus does not represent true tropical forest, but we are unable to determine how much. The FAO reports an annual deforestation rate in Cuba of 0.9% during the decade of the 1980s. We are, nevertheless, convinced that the forest cover has indeed increased since 1959. In the case of Puerto Rico, the island was almost completely deforested by the 1930's, where 99% of its primary forests were gone by that time and with an estimated cover of secondary forest of 10–15% (Lugo, 1988). It had increased its forested areas to 31% (284,000 Ha) by the late 1970s, (Birdsey and Weaver, 1982).

[20]Unfortunately terms such as sustainable development or ecological development, or ecodevelopment and a variety of others have now been adopted and coopted by the very agencies that have been promoting ecologically damaging development in the past, changing only the name of what they do.

8

———◆———

Biodiversity, Agriculture, and Rain Forests

Despite the fact that there exists great potential to find new products in rain forests, possibly even cures for diseases like cancer, it is not really these utilitarian notions that fire the public imagination. Rather, it is the amazing fact that rain forests contain almost all the biodiversity on the planet, as discussed in Chapter 2.

Even disregarding all the potential rain forest products, rain forests' possible use as carbon sinks,[1] and the many other useful things have been imagined for or from rain forests, observations about rain forest biodiversity cannot help but strike a deep chord. Although rain forests cover only about 7% of the surface of the earth, they are thought to contain more than 50% of its biodiversity.[2] Anyone would respond with deep emotion if they learned their home town — inhabited by their mother, father, brothers, sisters, aunts, and uncles — was about to be visited by a natural disaster. Perhaps a similar deep feeling is stirred when we note, as biological creatures, that the vast majority of our living relatives are in the rain forest, and that rain forests are being rapidly destroyed.

In a sense this is the ultimate concern. We must acknowledge all the potential uses of this biodiversity, but it is important to recognize that a deeper principle is involved. If some force was trying to destroy the world's art museums, all thinking humans would be concerned. Such a force is right now, destroying treasures equally important and perhaps even more irreplaceable than the contents of the world's art museums. Our concern is justified.

Biodiversity and Utilitarianism

Despite the above, there is no doubt that a utilitarian focus on biodiversity is to some extent valid. While the ultimate reasons for concern over the loss of biodiversity may not fall within the category of utility, the historical record and common sense both suggest that biodiversity indeed has value as a source of useful things for humanity.

The last five centuries certainly make this clear. Much of the initial economic push that led to the European conquest was the search for products from tropical lands. The European desire for spices drove the indefatigable search for the almost mystical spice islands, and accounts to some extent for the early Dutch experiments in colonialism. The English obsession with tea sweetened with sugar arises from their domination of colonies in south Asia and America. Recall the ultimate purpose of Captain Bligh's journey. Bananas, chocolate, coffee, tomatoes, and many other food and drug crops that today we take for granted, have their origin with entrepreneurs, pirates, and governments seeking to make use of tropical biodiversity. In fact, these forests contain many of the wild cousins of domesticated food and industrial plants. In addition, many modern medicines come from plant material extracted from the rain forest.

It would be folly to suggest that this historical pattern has suddenly come to an end. Indeed, as we write, pharmaceutical companies from Merck to Bristol-Meyers are searching the remaining patches of tropical rain forests for products they can market. The centerpiece of Costa Rica's National Institute for Biodiversity is a search for products that can be sold to pharmaceutical companies.

But there is another component to the biodiversity question that is sometimes lost in exaggerated rhetoric. When bird-watchers first come to a tropical rain forest area, they are frequently quite disappointed. The birds are difficult to see, largely due to the lush vegetation that provides them with cover. Thus one often sees the amusing irony of bird-lovers who, having visited a tropical country to see the beautiful birds of the rain forest, seek out whatever patch of open non-rain forest area they can find, in order to actually be able to see their birds. Indeed, in any rain forest reserve in the world, if you wish to spot bird-watchers, seek out the abandoned agricultural areas.

Furthermore, recall from Chapter 2 that disturbance, sometimes modest, sometimes catastrophic, is a perfectly natural component of the rain forest cycle. So the animals and plants of the rain forest do not require an untouched cathedral-like green mansion, but rather expect a mosaic of dark understory, light gaps, old landslides, tornado and hurricane tracks, and the like. There is a parallel to this concerning the various forms of human activity in rain forest areas. Some activities are so minor that most organisms in the forest are unaffected, while others are so dramatic that the recovery of the forest is delayed for a very long time. We categorized the types of damage in Chapter 6 according to rates of recovery. But what about the component all agree is the critical feature, biodiversity?

This problem is usually approached by asking, "What would happen to biodiversity if all the rain forests were cut down?" It seems like a straight-forward question, but it is grossly misleading. Indeed little is known for certain about what happens to biodiversity when logging occurs. The image fixed in most people's mind is devastation: thousands of species replaced by a field of tree stumps. But this is not the consequence of timber harvesting per se. The agricultural activities that almost inevitably follow a logging operation are far more devastating than the logging itself. Indeed since the forest is usually resilient to major damaging events, absolute loss of species as a direct result of logging is probably rare. The trees resprout, the birds hide in the resprouting trees, and who knows what happens to the insects and other smaller animals?

In fact we have little evidence that the simple act of cutting the forest down does anything whatsoever to biodiversity. Of course the nature of the logging operation itself is likely to have an effect, and certainly its extensiveness is also likely to have an effect. But from the few studies available,[3] we really must withhold judgment. It may well turn out that, despite the devastated physical appearance of a hectare of logged forest, little biodiversity loss actually occurs due to the *direct* effects of a logging operation itself.

This surprising observation leads to a point made earlier. It is what happens *after* the rain forest is cut down that matters. If you cut it down and put a city in its place, clearly you have changed the biodiversity for a long time; perhaps forever. But if you cut down a few trees and go away, the trees resprout, ecological succession proceeds in the light gap produced by the cut, and there is little evidence that biodiversity is reduced at all. There is obviously a gradient between the cutting of a few trees and the construction of a parking lot. The rate of recuperation of the forest will be extremely rapid at one end and extremely slow at the other. But more importantly, the loss in biodiversity will be small at one end and large at the other. We must now ask a serious question: exactly how much biodiversity will be lost through various modifications of the rain forest environment? Surprisingly, despite numerous emotional pronouncements, little is known to answer this question.[4]

Biodiversity and Agriculture

Recent focus on the preservation of biodiversity is based on the suggestion that more species are becoming extinct in the latter part of the twentieth century than at any time in history. While efforts to curb these losses have intensified in recent years, emphasis has been on the preservation of a few charismatic and conspicuous organisms, or on creating pristine environments, mostly within national parks and reserves. In fact such organisms are a very small fraction of the threatened biodiversity, and such habitats represent only a small percentage of total land area. While obviously, on a unit-area basis such habitats are the most biologically diverse in the world, it is also true that managed ecosystems are far more extensive in area. Since the combination of managed ecosystems and human settlements cover approximately 95% of the earth's terrestrial surface, it might be argued that it is as important to examine patterns of biological diversity in managed ecosystems, as it is in highly diverse unmanaged ones like tropical rain forests.[5]

For example, with 622,000 hectares (12.2% of the total land area) covered by preserves,[6] and a total of 27% of its land under some sort of protection[7] Costa Rica has one of the world's highest proportions of land under protected status. Yet even in this remarkable case, almost four million hectares (73%) are covered in agroecosystems, managed forests, and human settlements. Furthermore, while a fraction of that area is some form of traditional agriculture, the rest has been, or is being, transformed into high-input monocultural systems. Recently we have come to understand that indirect biodiversity losses through agricultural transformation in this system may be very large.[8]

Regarding biodiversity in agroecosystems the literature is clear on one major point: two distinct components of biodiversity are recognized. The first component is the biodiversity associated with the crops and livestock purposefully included in the agroecosystem by the farmer. A traditional home garden is more diverse than a modern wheat field because many crops are planted in the former and only one in the latter. This is *planned biodiversity*. While planned biodiversity is perhaps the most visually obvious component, and has received the greatest attention, *associated biodiversity*, is at least as important. This component includes all the soil flora and fauna, the herbivorous, carnivorous, and fungus-feeding insects, the birds and mammals, the associated plants (some of which are weeds), and more.

Most discussion of biodiversity in agroecosystems has centered on the function of the planned biodiversity. For example, it is widely believed that a diverse assemblage of crops in a multiple-cropping system reduces market risk. Trees in coffee plantations are considered necessary to provide partial shade for coffee plants. Rotation of soybeans with corn provides nitrogen to the soil. Many other examples could be cited.

Associated biodiversity is less well understood. When we compare a Javanese home garden with a pesticide-drenched rice paddy, we find a rich diversity of soil micro and macro flora and fauna in the former, but not the latter. This is the associated biodiversity. While the function of this component is debatable, whatever it may be, it is affected indirectly through the planned biodiversity of the home garden.

We illustrate this relationship in figure 8.1. Planned biodiversity has a direct function, as illustrated by the bold arrow connecting the planned biodiversity box with the ecosystem function box. Associated biodiversity also has a function. But associated derives from planned. Thus, planned biodiversity also has an indirect function, illustrated by the dotted arrow in the figure, which is realized through its creation of the associated biodiversity. For example, the trees in a home garden create shade, which makes it possible to grow some sun-intolerant crops. So the direct function of this second species (the trees) is to create shade. Yet along with the trees might come small wasps that seek out the nectar in the tree's flowers. These wasps may in turn be the natural predators of pests that normally attack the crops. The wasps are part of the associated biodiversity. The trees, then, create shade (direct function) and attract wasps (indirect function).

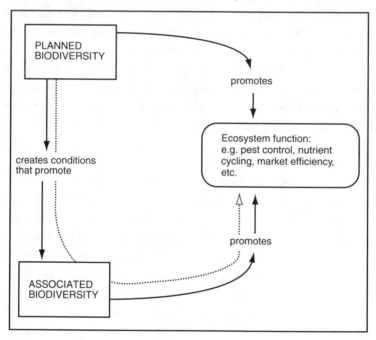

Figure 8.1. The relationship between planned biodiversity (that which the farmer decides to include in the agroecosystem) and associated biodiversity (that which moves into the agroecosystem after it has been set up by the farmer), and how the two promote ecosystem function.

This simple example notwithstanding, the actual rules that dictate how biodiversity is translated into function are different in managed and unmanaged ecosystems. Farmers not only plan the biodiversity, but they engage in management practices that may very well alter the manner in which that biodiversity is translated into function, either the direct or indirect.

How associated biodiversity is determined by planned biodiversity is a legitimate and interesting question in its own right, independent of the ultimate function of the associated biodiversity. Yet a different set of questions arises from consideration of the continuing changes in human agricultural activities. As farmers change their management practices, they may or may not change the planned biodiversity. Clearly as the old native multiple-cropping system of Mexico is replaced by monocultures of corn, we expect changes in the associated biodiversity also. But even a change in corn monocultures in Central America from a high labor/low capital input system to a system involving high capital outlays also has a potentially important effect on associated biodiversity. Simply the use of pesticides may result in an enormous biodiversity loss. Thus, changes in husbandry practices may force changes in the way planned biodiversity functions, the way it translates into associated biodiversity, and in the way the overall biodiversity affects ecosystem function.

Given these observations, we can summarize the relationship between agricultural intensification and associated biodiversity as a simple graph (figure 8.2). While there is controversy over what actually constitutes intensification, there is general agreement about the extremes. At one end, for example, are the traditional Javanese home gardens, while at the other are the modern rice plantations — paddies prepared with machines, directly seeded with automatic seeders, sprayed with pesticides from airplanes, and mechanically harvested. At one end we find the small patches of corn and beans that Native Americans used to plant in forest openings in Michigan, while at the other end we find the modern wheat fields of Nebraska or the Ukraine. It is not difficult to recognize the extreme cases, a patch of corn and beans planted in a natural light gap at one end and a banana plantation at the other. While most agree that biodiversity decreases as intensification increases, the exact form of that pattern elicits significant discussion and debate. In figure 8.2 we have illustrated what we believe to be the four qualitatively distinct possibilities.

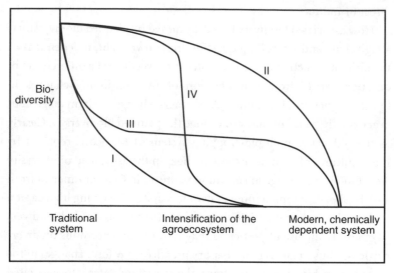

Figure 8.2. How biodiversity changes as a function of the intensification of the agroecosystem, with the four distinct patterns as discussed in the text.

Most ecologists concerned with biodiversity in undisturbed ecosystems have tacitly assumed that the pattern exhibited by curve I is the most likely case: a dramatic loss in biodiversity as soon as any human use and management is brought to bear on the ecosystem. The other extreme, curve II, would perhaps seem unlikely to most ecologists, yet we know of no data or theoretical argument that speaks to the issue. Indeed, the small amount of data available,[9] might be interpreted as support for a type II curve. However, we suspect that in most cases something between these two states will be the truth; either curve III or curve IV.

Curve III is a softer version of the ecologist's expectation and simply indicates that, after an initial dramatic loss in biodiversity, the further loss as management intensifies is relatively slight, until the extreme of the truly modern systems. Curve IV is perhaps the most interesting case if one is concerned with biodiversity preservation per se. This curve suggests that initial stages of management do little to the overall biodiversity, the loss in biodiversity remaining gradual until some critical stage of management intensity. At that critical stage biodiversity declines very rapidly. If biodiversity per se is the concern, it is obvious that planning activities ought to be focused on maintaining management intensities below that critical point, rather than on aiming at a zero-management strategy. This would mean focusing on sustainably-managed agroecosystems rather than pristine unmanaged ones.

Concerns about loss of biodiversity have usually centered on the transformation from natural forest to agriculture, while the transformation from low- to high-intensity forms of management, which today involves an enormous amount of land, has been practically ignored. It is clear that the greatest biodiversity per unit area exists in tropical forests, and since these forests are being destroyed at such a rapid rate, the bulk of the world's efforts at cataloging and conserving biodiversity are justifiably aimed at these disappearing ecosystems. Yet it seems equally obvious that if, as is sometimes suggested,[10] there exists substantial biodiversity in traditional agroecosystems, then transforming them into modern, intensively managed systems represents a potentially significant loss of biodiversity. Surprisingly, we know as little about biodiversity in traditional tropical agroecosystems as we know about biodiversity in the tropical rain forest, and little idea which curve in figure 8.2 may actually be correct. With a formidable amount of land in the tropics currently undergoing transformation from traditional to modern agroecosystems,[11] it could well turn out that substantial biodiversity is being lost through this change. Clearly this warrants further study.

Certain kinds of biodiversity attract the greatest attention. Little sparks the public imagination more than elephants or other *charismatic megafauna*. Ridicule was heaped on conservationists in the United States when they voiced concern about the extinction of the snail darter, but no one takes lightly the potential extinction of the African elephant. We agree with those who assert that focus on charismatic megafauna helps attract the attention of a public as yet unaware of the problems associated with biodiversity loss in general. There are, however, also problems with this focus. Little in the popular conservation literature suggests that initial concern for charismatic megafauna ultimately translates into any other concerns, indeed the evidence might actually support the opposite.[12] If all available political energy becomes focused on such creatures, little energy remains for concern about the rest of the world's biodiversity.

Small organisms such as insects are frequently more specialized and thus are especially susceptible to extinction when vegetation and habitats are modified. Small organisms also represent the vast majority of the world's biodiversity. For example, there are almost 200,000 described species of arthropods,[13] and it is estimated that there are ultimately anywhere from five million to thirty million species of insects alone. In addition, arthropods are potentially important pests, as well as the natural enemies of pests in agriculture. Other groups of organisms are likewise enormously diverse: plants, roundworms, fungi, algae, bacteria, and more. The smallest and least noticeable of organisms constitute almost all of the biodiversity on earth. Nevertheless, most popular concern with biodiversity loss is aimed at the charismatic creatures, the elephants, tigers, rhinoceroses, and the like. For the very same reasons that we should focus on agroecosystems when concerned with biodiversity loss, our "organism" focus should aim at the little ones.

An Example: the Coffee Agroecosystem of Central America

The significance of the above observations can be recognized in almost any managed ecosystem in the world. An example in which the problem has been recently evaluated is the coffee agroecosystem of Central America. It is a tropical habitat found over a large area of the world; in its traditional form it has a forest-like structure; for the past decade it has been undergoing transformation from forest-like to monoculture; it harbors a range of insect communities; and it displays obvious patterns of biodiversity loss in vegetation, implying similar loss patterns in the insect communities.[14] In a sense, this habitat represents a natural experiment in biodiversity dynamics. Replicated experiments have effectively been set up with some coffee plantations retaining the presumed biodiversity-preserving traditional form and others in some stage of "deterioration" as a result of the "modernization" process.

As with other major ecosystem transformations in tropical latitudes, the transformation of the coffee agroecosystem involves spectacular landscape changes. The traditional system follows the common pattern of agroforestry, with a variety of shade tree species frequently interspersed with fruit trees, sometimes with relatively dense plantings of bananas in a forest "canopy" above the coffee bushes. The coffee itself is managed at the level of individual coffee plants. Pruning creates small light gaps into which cassava, yams, or other annual crops are planted. When a whole group of coffee bushes are to be renovated (removed and replanted with new bushes), a larger light gap is created and may receive a planting of corn, beans or other light-demanding crops. Thus, traditional coffee farms share many structural attributes normally associated with forests (figure 8.3).

The "modern" monocultural system that is being promoted all over the world could not be more different. All the shade trees are eliminated, the traditional coffee varieties are replaced by new sun-tolerant and shorter varieties, which are genetically homogeneous and pruned either by row or by plot, and are heavily dependent on agrochemicals, especially herbicides and fertilizers (figure 8.4). These two systems represent the two extremes in a continuum of management systems with varying degrees of complexity.

Although such transformation is qualitatively spectacular at the landscape level, little quantitative data have been collected to estimate biodiversity loss. The entomological laboratory of the National Autonomous University of Costa Rica[15] has been doing studies of the arboreal arthropod fauna and has encountered some fascinating results. Arthropods were collected from shade trees, coffee bushes, and the ground. Samples were taken in 1) a traditional coffee farm characterized by a high diversity of shade trees, 2) a more intensively managed plantation, and 3) a "modern" monocultural plantation. These sampled farms represent three points along the imagined intensification gradient of figure 8.2. As expected, the biodiversity of the arthropod groups declined as intensification gradient increased, as shown in table 8.1. But more importantly, the diversity of arthropods in the traditionally managed system was impressively high. Indeed, it was almost within the same order of magnitude as that reported for beetles and ants living in the canopies of rain forest trees.[16]

Figure 8.3. The traditional coffee agroecosystem system in Central America, aerial view (top) and view from within (bottom). Photos from the Central Valley of Costa Rica.

Figure 8.4. The modern, chemically-intensive, coffee agroecosystem in Central America. Entire landscapes are covered with monocultures of coffee (top and middle photo). Bottom photo shows worker, with very little protective gear, receiving pesticide in his backpack sprayer. Photos from the Central Valley of Costa Rica.

Other studies have corroborated this basic tendency: the associated biodiversity follows the same pattern as the planned biodiversity. As the system is transformed from its traditional state — three or four species of shade trees, some twenty to thirty species of fruit and lumber trees,

plantains, bananas, and understory crops, and light gap annual crops —
to a monoculture of coffee bushes, the associated number of beetles,
ants, wasps, and spiders decreases. In this example it appears that either
the type II or type III curves of figure 8.2 are the true patterns. Yet even
in this system — probably the best-studied agroecosystem thus far —
from the point of view of biodiversity changes, we are not certain what
the true quantitative effects are.

Table 8.1. Insect species found in the three coffee systems.

Position along intensification gradient (see figure 8.2)	Beetles in shade trees	Ants in shade trees	Wasps in shade trees	Ants on ground	Beetles on coffee bushes	Ants on coffee bushes	Wasps on coffee bushes
Traditional	128	30	103	25	39	14	34
Intensively managed	50	5	46	14	29	9	31
Modern agrochemical system	0*	0*	0*	8	29	8	30

*There are no shade trees in the modern system.

Models for the Future

The above implies an alternative model for the preservation of biological diversity in rain forest areas. The classical model has been to set
up preserves by purchasing the land on which the rain forests sit, and to
declare such land inviolable for other human uses. The preservation of
biodiversity is viewed as the antithesis of other types of human activities,
such as forestry and agriculture. The classical model thus strikes a
Faustian bargain in which the inviolable lands are to be religiously
respected and all other lands are to be ignored — let us have our 100
hectare wilderness and we will remain silent when the other 100,000
hectares of forest are destroyed.

Our alternative model takes into account what we already know about biodiversity, not only in tropical rain forest areas, but also in the managed states: the agroecosystems and forestry systems. There are two principles upon which the alternative model is based. The first principle is that significant amounts of biodiversity occur in non-rain forest areas like forestry and agricultural areas. However, a corollary to this principle, as illustrated by the example of the coffee agroecosystem, is that specific types of forestry and agriculture imply varying degrees of biodiversity. The second principle is that biodiversity preservation should be part of the planning process in agricultural and forestry areas. If, for example, we had known ahead of time the devastating consequences of DDT with regard to biodiversity, its introduction as part of the agricultural system would not have been permitted in the first place. Similarly, the introduction of a new coffee technology should not be judged solely on the basis of the pounds of coffee beans it produces per hectare this year, but also on the basis of its impact on the biodiversity of the areas in which it is to be introduced. Just as zoning laws sometimes prohibit particular activities in certain areas of cities, zoning laws in potential agricultural and forestry areas ought to be enacted. One consideration for these laws should be the impact of any activity on biodiversity.

While we do not imagine that such laws will be enacted in the near future, we hope that as consciousness is raised concerning the loss of biodiversity in forestry and agricultural areas within rain forest habitats, movements in this direction will evolve. We return to this political point below, but first let's explore what this alternative model might look like.

In figure 8.5 we illustrate a typical graph of the number of species plotted against area. The area may be just the size of a sample in an region, or it may be the size of a preserve, or group of preserves. The number of species will tend to rise as the area increases, but eventually level off at some specific value. We have illustrated a situation in which there are a total of 100 species in the entire region.

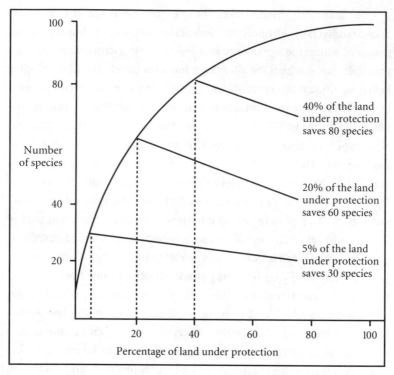

Figure 8.5. Biodiversity (number of species) preserved within protected areas as the percentage of land under protection increases.

Note how the number of species we can expect to be preserved drops as the area devoted to preserves decreases. With 40% of the land under reserve status, we save 80 of the 100 species. With 20% in reserves we save only 60 species and the number of species saved drops to 30 if we preserve only 5% of the land. This is a well-known relationship — the number of species increasing with increasing area — and has led many to suggest that for maximum conservation of biodiversity conservation the size of biological preserves must be as large as possible. This seems so obvious. But is it?

If we back up and ask about the land outside of preserves, and take that land into account in our calculations, the picture begins to look quite different. If we assume, as many conservationists are willing to do, that close to 100% of the species will go extinct outside of the reserves as modern agriculture continues its assault on nature, yes the analysis is true that the larger the reserve the better. But if we concern ourselves with the management of the land outside of the reserves, it is not at all clear what the correct strategy might be. For example, if we put into place regulations that require no more than a 50% loss in species diversity in those areas under management, putting 20% of the land in reserves will save an equal number of species as would 40% (the 60 species preserved in the reserve plus 20 of the remaining 40 species that do not go extinct because of the regulation). We display, for the example in figure 8.5, the expected biodiversity preservation as a function of various management strategies in table 8.2.

Table 8.2. Hypothetical biodiversity preservation (out of a total of 100 original species) under various management strategies.

Size of reserve as percent of total area	Allowable extinction outside of reserves		
	10%	50%	100%
40%	98	90	80
20%	94	80	60
5%	84	65	30

This table makes it clear that an overall strategy that includes concern with what happens outside of the biological reserves is more likely to conserve the greatest biodiversity. In the real world it is not likely (nor desireable) that biological reserves will encompass all of a region, and the most that can be expected is something more along the order of 5% of the land in reserves (Costa Rica, for example, is one of the world's largest preservers, with 12.2% of the national territory in preserves). But even with this small quantity, a great deal of biodiversity will be preserved if we simply concern ourselves with the loss of biodiversity in the managed ecosystems outside of the reserves. In the above example, with only 5% of the land under formal reserve status, but with the requirement that management activities not reduce biodiversity by more than 10%, more species are saved than in the case of 40% preserved in pristine islands amidst a sea of decimation.

Unfortunately we know too little of biodiversity in general, or specifically of biodiversity in managed ecosystems like agriculture and forestry, to be able to judge positively which model, the classical or the alternative model, will preserve more biodiversity. But given preliminary studies, and given the poor record of modern agriculture in biodiversity conservation, we strongly suspect that our alternative model is more appropriate. In the coffee agroecosystem, for example, biodiversity of arthropods appears to be on the order of 60–70% of what the original forest might have held. Converting that agroecosystem to its highly capitalized modern form reduced that biodiversity to perhaps 5% of the original.

The alternative argument is sometimes made that the modern agroecosystem is sufficiently productive on a per hectare basis that a smaller amount of land can be put into agricultural production in order to satisfy human demand, thus leaving the rest for biodiversity conservation. Putting aside the abominable record of "modern" agriculture in terms of satisfying human needs,[17] we must ask how much biodiversity is lost in the transformation to the modern state, and how much is preserved by keeping the rest in a pristine state? This is not a simple issue, and indeed we know too little about biodiversity in managed ecosystems to give a proper assessment. But again, we strongly suspect that a terrain of biological devastation dotted with small islands of pristine rain forest will preserve far less biodiversity than a landscape of sustainable agricultural and forestry activities surrounding a smaller number of pristine areas.

However appealing our alternative model might seem, political realities are something different. Consider again the coffee agroecosystem. In the early 1990's the price of coffee bottomed out, mainly due to post Cold War political maneuverings. The problem was overproduction on a worldwide scale, a fact agreed upon by economists, planners, and coffee farmers alike — although coffee agronomists continued to insist on a religious commitment to increase production. Most coffee-producing countries formed a new cartel in 1992 and prices began rising in 1993 as a result of the members agreeing to retain significant quantities of coffee off the market.[18] So we have a dual problem. First there is a biodiversity crisis, and second there is an overproduction of coffee. If it is true that a significant amount of general biodiversity is sequestered in traditional coffee plantations, it would make sense to try to preserve them, which means preserving the traditional way of coffee production. But, the argument goes, the traditional way is not very productive, that is, the amount of coffee produced per area is much lower than in the modern system. Yet since all agree that the major problem with coffee is overproduction, why should we care about producing more per hectare?

There is a "tragedy of the commons"[19] problem here both nationally and internationally. Individual farmers try to produce as much as possible but in so doing they collectively saturate the market and prices are consequently reduced — a tragedy for all. But, as an ideal solution, why not give "biodiversity credits" to those farmers who decide to return to traditional production, rather than subsidize farmers to keep coffee off the market? Investing government funds in this sort of conservation would treat two problems simultaneously: the loss of biodiversity through agricultural modernization and the overproduction crisis in coffee.

Again the argument could be made that putting only a few farms in high-yield production and letting the rest revert to forest may preserve more species and reduce production so as to raise prices. We repeat, not enough is known about the patterns of biodiversity in agroecosystems to say which of these two strategies would preserve the most biodiversity. Our prejudice is that an entire landscape devoted to sustainable organic coffee production, using shade trees, fruit trees, plantains, and understory crops as part of the system, is likely to preserve more biodiversity than islands of forest in a sea of modern pesticide-drenched agriculture.[20] And, to repeat a theme of previous chapters, who will benefit from a few highly technified coffee farms, and what will the rest of the people do for a living?

It is on this note that we must end our discussion of biodiversity per se. In a previous chapter we have already devoted considerable space to the landscape model we propose. We can only add, recalling our earlier argument, that even if biodiversity is not more completely preserved in our landscape model, there are social and political reasons why this model is the preferred one, perhaps even the only viable one. But we must emphasize, our landscape model is simply not possible without land security.

---

[1] As part of the normal process of photosynthesis, trees absorb carbon dioxide. If they are cut down, all of the carbon contained within them will be released into the atmosphere, thus contributing to greenhouse gases and global warming.

[2] It is estimated that approximately 80% of all insect species, 60% of all plant species, and 90% of all primate species live in rain forests, (Westoby, 1989).

[3] Johns, 1991; Karr, 1971; Michael and Thronburgh, 1971; Lovejoy et al., 1984; 1986.

[4] Swift et al. (1994) note that as agroecosystems are transformed the pattern of biodiversity may take many different forms, and we are largely ignorant which of these forms actually occurs in nature.

[5] Pimentel et al. 1992.

[6] Boza, 1993.

[7] This includes protected categories such as forest reserves, wildlife preserves, protection zones, biological reserves, national parks, and recreational areas.

[8] Swift, et al., op cit.

[9] For example, several studies suggest that insect biodiversity is larger in systems where more than one crop are grown at the same time, that is, where planned biodiversity is higher than in the normal modern monoculture but, of course, much less than in the natural ecosystem that it replaced (Risch et al., 1983; Andow, 1991). Furthermore, considering the organisms that live in the soil, their biodiversity pattern may very well show a type II response (Swift et al, 1979).

[10] Gliessman, 1993; Dover and Talbot, 1987; Altieri et al., 1987; Paoletti, 1988.

[11] Structural adjustment programs designed by the World Bank and IMF are likely to accelerate this process of agricultural intensification. In addition, the FAO's global agricultural strategy for the future focuses on agricultural intensification.

[12] Bonner's study of the conservation community concerned with the African elephant suggests that the drive for ever greater contributions may actually cause this sort of concern to even overlook the fact that the species of concern is harmed by fund-raising tactics. Bonner, 1993.

[13] Arthropods include insects, spiders, crabs, lobsters, and other such creatures with jointed legs.

[14] The data on arthropod diversity in coffee ecosystems is from our studies in Costa Rica (Perfecto and Snelling, 1995; Perfecto and Vandermeer, 1994).

[15]See footnote 13. These studies are cooperative ventures between the National Autonomous University of Costa Rica and the University of Michigan.

[16]Erwin and Scott, 1980; Wilson, 1987; Adis et al., 1984.

[17]See for example Shiva, 1993.

[18]Significantly complicating this picture is the freeze that hit Brazil in 1994. Coffee prices soared as news of coffee plantations being destroyed by the unexpected cold wave in Brazil reached the world.

[19]The meaning of the phrase here is obviously different than its original intended meaning (Hardin, 1968), but the economic structure is precisely the same. The market is the "commons" and each individual producer is trying to take as much out of it as possible, leading to over supply and dramatic declines in market prices, a tragedy for the coffee farmers.

[20]See Shiva, 1993.

9

◆

Who Constructs the Rain Forest?

In a beautiful rain forest preserve in the Caribbean lowlands of Costa Rica, one of our students recently remarked, "The trouble with this place is that even on the back of the property you can hear trucks on the road." You might ask what trucks, or the absence of their sound, have to do with a tropical rain forest. The answer is nothing and everything. To this student, a certain feature of the tropical rain forest was the solitude, the isolation, the feeling of being removed from the modern world. His point of view is similar to many who have spent time appreciating the natural world. Nature is what was there before human systems came to dominate and alter it. Furthermore, nature is always seen as more beautiful than its human-modified form. This idea is a social formation that we call nature worship.[1]

At another extreme, the Ston Forestal company is planting huge monocultural plantations of Melina, an introduced tree that grows very rapidly. To take advantage of substantial tax breaks offered by the Costa Rican government for "reforestation," Ston does not call its operation a tree plantation, but rather a reforestation effort, and the result of their effort will be a rain forest, by legal definition. Clearly this is a totally different concept of a rain forest, probably dictated more by the desire for tax credits than any honest feeling about tropical rain forests.

Looking back in history, consider the following description of a tropical rain forest:

> ... broken only by the clanging clamor of the torrents, the growling of tigers, and the swarming of infinite vipers and venomous insects....the plague of vampire bats ... extends until Brazil, treacherously sucking in the hours of dreams the blood of men and animals. There also, by the side of the Brosymum [tree] ... grows the Rhus juglande ... whose mere shade puffs up and scars the careless wanderer. There one begins to suffer the privations and calamities of the wilderness, that ... grow so as to at times make of life there scenes whose horror could figure in the Dantean pages of Purgatory and Hell.[2]

A more recent assessment suggests that the tropical rain forest is,

> ...in a class of its own. It is akin to other super-size spectacles on Earth, such as the Grand Canyon and the Victoria Falls. However much you read about the scene, however many photographs or films you see of it, nothing prepares you for the phenomenon itself with its sheer scale and impact. You gaze on it and you feel your life has started on a new phase. Things will not be the same for you again after setting eyes on something that exceeds all your previous experience.[3]

So what, in point of fact, is a tropical rain forest? Is it a hell, or a paradise? Is it a place to be cleared for a farm, or a place for birdwatching? Is it somewhere to escape from the modern world, or a collection of Melina trees? Is it something that contains cures for cancer, or an impediment to cattle, a source of poisonous snakes, agricultural pests, or enemies of agricultural pests? Clearly there are at least two components of the tropical forest: 1) the trees and other plants, soil, insects, animals, and how they are all related to one another, and 2) the observer, who may be a local farmer, a foreign industrialist, an ecotourist, or something else entirely. In short, there is text and reader (1 and 2, respectively), and the nature of the tropical forest cannot be understood as anything other than the relationship between the two.

Our use of the terms *text* and *reader* is intended to call attention to the relationship between recent theories of literary criticism and the notions of nature that we humans have. Contemporary literary theorists tend to view various texts in the context of who is reading them. It is impossible, they argue, to analyze a text independently of its context, and its most important immediate context is the person engaged in reading it. True meaning, then, is not to be found in the text alone, but rather in the *reading* of the text, both the actual words and the context in the reader's mind.[4]

This view easily reconciles the dramatically divergent views of the tropical forest with which we opened this chapter. The text, so to speak, is the physical elements in the area called forest, and the reader is he or she who observes that text. For our student, the forest was a pristine wilderness. In reading the text, what stood out in his mind was the sound of the trucks on the nearby road. For the economists of Ston Forestal, the forest is read as a bunch of trees waiting to be harvested for profit. For the Spanish explorer searching for gold mines, a reading of the text views the nature of the forest as akin to, "Dantean pages of Purgatory and Hell." For the contemporary First World conservationist, reading the text constructs the nature of the forest as something that makes your life start on a new phase. What then *is* the nature of the tropical forest? It is as much a social construction as the Pyramid of the Sun in Mexico, the Holy Trinity, or the peculiar notions of democracy that western nations have constructed.

If the true nature of a tropical rain forest is as much a social construction as other human structures, both physically and conceptually, it may be "deconstructed" and "reconstructed." Is this not what our student did when he suggested that a true tropical rain forest is not simply the plants and animals that make up the physical aspect of the forest, but includes the lack of truck sounds? Did not Ston Forestal deconstruct the romantic notions of what forests are and reconstruct, largely for the purpose of tax credits, the forest as nothing but a stand of trees? And the Spanish explorer and conservationist likewise have constructed their own notions of the rain forest. The true nature of the rain forest is not to be found in the physical aspects of the forest alone (the text). As we view those physical aspects (as we read the text), we construct our own nature of the rain forest. If this is true, we must ask whose reading of the rain forest represents truth?

This sort of reasoning can easily lead to a quagmire of excessive relativism — if everyone can have a different reading of the same text, there is no true meaning. It is most definitely not our intention to promote such a view. While it is imperative to deconstruct our notion of nature, probing the social relations that give the concept meaning in the first place, this in no way represents a license. That nature is a human construction does not imply a crude relativism, let alone a license to destroy, but rather cries out for more cautious analytical thought. Nature should be viewed much as a piece of literature, as a phenomenon to be analyzed, not as a religious icon. Just as art can be beautiful, or ugly, nature can likewise take on such qualities.

The current assumption of some conservationists that pristineness and uniqueness are the primary criteria for a proper appreciation of nature derives from a kind of romantic essentialism. It probably should be abandoned. Rather, the aesthetic sense of nature should be continually evaluated and reevaluated, much as we do with literature or art. Pristineness and uniqueness certainly enter the equation, but not necessarily as the dominant features. Similarly, making profit potential the primary criteria for a proper appreciation of nature derives from underlying greed and should also be abandoned. Rather, the potential for nature to make a profit for someone should be continually evaluated and reevaluated, and should include the question of profit for whom?

If the nature of the rain forest is to be adjudicated as a social construction, part of the evaluation of our particular construction ought to be the question, "Who is doing the constructing in the first place?"

The Nature of Preserving Nature

Nature is not merely the venerable background in which we live, but rather is one component in a dialectic of which we are the other. The environment of any organism, including ourselves, exists only in the context of that organism. The force of gravity is of little import to a bacterial cell, whereas a monkey swinging from branch to branch must be constantly aware of it. In this sense environment (and therefore nature itself) is distinct from the form in which it is popularly conceptualized. Frequently it is thought that an organism lives in a metaphorical landscape that contains peaks and valleys. The organism is continually ascending the peaks (called adaptive peaks) through the process of evolution. Such a landscape is referred to as an adaptive landscape and is a favorite pedagogical tool for teaching evolution.[5]

However, modern biology has been rapidly altering that concept in recent years. An organism's environment is no longer thought of as a fixed landscape, but rather an ever-changing one, responding to the actions of the organisms that continually act on it. An adaptive peak for a beaver is the beaver pond, but the beaver creates the beaver pond in the first place. Indeed, the metaphor of a landscape filled with peaks and valleys should be replaced with a trampoline whose contour pattern adjusts to the changing organisms which metaphorically bounce on it.[6]

Similarly, the part of the environment we humans call the natural world, or nature, is trampoline-like, shifting in time and in response to various readers.[7] The rain forest was a valley for the Spanish explorers, but is a peak for the romantic conservationist. The essence of nature is thus not only a combination of text and reader, but a negotiated conceptualization involving a variety of readers, each of whom, necessarily, has only a partial reading. And we can fully expect that conceptualization will change as different readers engage with the text and interact with one another.

Given this view, we see two distinct models. The first gives clear priority to particular readings, the second sees the readings as emergent properties of shifting alliances. The programmatic consequences of each model are distinct. The *prioritized particular reading* model offers the conclusion that political liberation is a *prerequisite* to nature conservation, while the *shifting alliance* model concludes that nature conservation is a likely *consequence* of liberation. From a practical political point of view, the first model suggests that the goal of preserving tropical rain forests is not possible before poverty has been eliminated, and thus has certain pragmatic appeal for the political activist. The second model suggests something far more complicated, that the notion of preserving the tropical forest itself must be continually reanalyzed in line with extant political realities.

First, let us deal with the particular reading model. The view that a pristine nature exists outside of the experience of human beings is to some extent romantic idealism, perhaps stemming in part from escapist tendencies, which result from the problems of the modern world. An alternative is to view nature's beauty as similar to that of a poem or work of art. It is the appreciation of human beings that makes it beautiful. Nature, in this sense, is as much a product of human construction as a Frank Lloyd Wright building, the pirouette of a ballerina, the solution of a complex equation, or the beat of a punk-rock band.

One of the consequences of the latter view is that, much as is argued for the capacity to appreciate fine art or music, only "educated" humans can appropriately appreciate nature. The implications of this view are not as elitist as would seem at first glance. Rather it acknowledges the appreciation and knowledge, which come from education as a human right. Landless peasants will appreciate the aesthetics of a rain forest differently when they are no longer forced to view it as a potential farm, and, rather, can view it as a source of inspiration and potential knowledge. Of course, today's intellectuals may very well view the rain forest differently in a world where former peasants hold equal political power. The former peasant farmer with a Ph.D. is unlikely to read the rain forest text as an impediment to his family's betterment, as he is forced to do today.

It is tempting to go even further and formulate a program, at least in theory, for the "correct" construction of nature. Our own construction is a complex mixture. John's construction of nature arises from (at least)

Tarzan movies, landscape paintings, boy scoutish nature lore, summer walks in the north woods with his mother, fishing with his father, adventure movies like *King Solomon's Mines*, teaching ecology at the university, personal relationships with others who love the outdoors, sharing visions of pristine wilderness with friends and acquaintances, and so on. Ivette's construction arises from (at least) weekend excursions to the El Yunque rain forest, the beaches of Guanica with their pollution from petrochemical companies, television nature shows, the mystery of the ocean and its fauna, visits to her uncle's farm, lizards on the walls of her house, teaching tropical ecology at the university, and more. Obviously both are special visions, despite their feeling of generality. It is difficult for each of us to imagine that our view may not be the "correct" construction. But if our previous argument holds, there is no clear way to prioritize our construction over the construction of the peasant who must cut down the forest to feed her family. It gives the economist at Ston Forestal as equal a reason for arguing his view as it does our student for arguing his.

We now turn to the shifting-alliance model. We suggest that the issue of nature construction may be brought into focus by an analysis of power structures in society. If nature is socially constructed, similar socio-political conditions will produce similar social constructions. Capitalists seek to construct nature in a way that will maximize profits. Thus, a poisoned river is not necessarily a negative in the capitalist's construction. Peasant farmers seek to construct nature in a way that will allow them to survive. Thus, a deforested patch of land is not necessarily a negative in the peasant's construction. Workers seek to construct nature to preserve employment conditions. Thus global warming is not necessarily a negative in the worker's construction. The nature of the construction of nature is based in part on one's power position in society.

One's construction of nature will be specific to one's history, and all constructions will necessarily be partial since all experience is partial. As various alliances form, the constructions will consequently change. As alliances form for political purposes, different constructions will follow. Those alliances that achieve a share of political power will contribute more to the construction of nature than those alliances that fail. Thus the construction of nature will be ever a partial construction based on shifting political alliances. The ultimate construction, achievable only with complete liberation, might look something like our contemporary preferred constructions. But it could also be quite different. However, if the construction of nature is inevitably a consequence of shifting political alliances, the nature of its intended conservation likewise will emerge from those political alliances. This makes it unpredictable.

Strategies for Rain Forest Conservation

In the same sense that one cannot envision the exact nature of an egalitarian society, yet can actively engage in the struggle to attain such a society, one may not see exactly how to construct the nature to be preserved, yet can actively engage in the struggle for a progressive conservation of nature. The theoretical principle of egalitarianism is clear enough, and the theoretical principle of conservation is clear in the same sense, but just as the slave society of Athens was regarded as egalitarian by the non-slaves, the conservation of today's conservationists may be (we think it likely) unrelated to a future construction. The construction of nature as created by today's educated classes (including parts of our own construction) may bear little resemblance to constructions to be elaborated in the future. And there is little question that future constructions will be formulated by different political alliances than those which exist today.

To the extent that our social construction of the rain forest is distinct from that of a local peasant farmer, or a banana company executive, or an indigenous person living in the forest, such distinctions will tend to diminish as society evolves into a more egalitarian form. Other differences may emerge, but those based on class must disappear as classes disappear. When the experiences of all people are constrained by the same material forces, social constructions of nature will likely be far less variable than they are currently. Most would agree that the utopian future of an egalitarian society is (alas) hardly on the horizon. Yet some tentative speculation may serve to initiate the debate on what might eventually constitute a "correct" construction of nature. In that spirit we offer a very brief speculation here.

The construction of nature that sees much of the natural world as being worthy of preservation, reconstruction, rational use, etc., has become popular in the West due to the thoughts and writings of privileged classes of people, mainly from the Developed World. It reverberates in so many ways with the reported constructions of many indigenous people, at least those less tainted by the ideological assault of western capitalism. Yet it is frequently at odds with the masses of poor peasants and slum dwellers of today's Third World. The poor who struggle to make ends meet in the global system have little time to contemplate the philosophy of biodiversity preservation when daily survival is at the top of their list of concerns. Their ancestors most likely shared a spiritual connection with the natural world that looked much like the worthy goals of contemporary western conservationists.

Conservationists from the privileged classes of the West and many indigenous people share certain elements in their vision of nature. Apart from utilitarian concerns that saving nature will find new cures for cancer, or help stop global warming, there is the honest argument that, "I wish to save the rain forest (or coral reef, prairie, or desert), because I find it immensely aesthetically pleasing, that cutting down rain forests would be like whitewashing the frescoes of Diego Rivera." This argument is, we believe, the motivation for the vast majority of First World conservationists, utilitarian claims notwithstanding, and bears a strong resemblence to the spiritual values of many indigenous people. Such views of nature result from particular readings of the text. If these particular readings are of value (and we believe they are), it makes sense to promote them. Yet the only way to promote a particular reading is to afford all potential readers the same reading privileges. But the peasant farmer struggling to feed his family is not likely to construct nature in the same way. In the end, the social construction of a nature that is beautiful and worth saving is not likely without the reorganization of contemporary structures of political power. In short, it will require the elimination of the situation in which some people are forced, by their class position, to read a negative text as they participate in the social construction of nature, to see nature as something to be eliminated, not venerated. If our general view of the appreciation of nature is even partially correct, our promotion of that view needs to take the form of a struggle for social justice.

Social Justice and Rain Forests

The image of peasant farmers cutting down rain forests has been a potent one in the past. In a very crude form it is the image that drives people to a variety of false analyses, such as overpopulation. Indeed, in the example of the *Sarapiquí*, we have heard respected scientists claiming that the problem in the area was simply the neo-Malthusian problem of too high a birth rate amongst the local peasants. Our position is clearly at odds with this point of view. Yet we too gain certain inspiration from our interactions with, and observations of peasant farmers. While the assumption that convincing peasants to have fewer children will help with the problem of deforestation is rather silly, the observation that peasant farmers do in fact cut rain forests down is accurate. We share a certain subtle conviction with the neo-Malthusians in that we feel there are too many poor peasant farmers. Our solution to the problem is, however, quite different.

We have gone to great pains to demonstrate that peasant farming activity, while directly responsible for much deforestation, is itself a consequence of something else. Thus, to engage in activities aimed at stopping peasant deforestation, it is necessary to aim at the causes, which are fundamentally structural. The politico-economic structures that define the way the Third World functions, are the structures that inexorably give rise to the peasantry to start with. The question of why the rain forest comes down may have the proximate answer to make way for peasant agriculture. But one must be studiously myopic not to be led to the next question, "Why are there landless peasants that need to cut forest?" This does not have as simple an answer, yet that answer is the key to intervention in the problem of deforestation.

Furthermore, to ask why there are landless peasants is much the same as asking why there is poverty, which is, at its most basic level, the fundamental question of progressive politics. The movement for social justice has represented a sedulous challenge to inequitable socio-political structures as far back as history records. And here we see the connection. If, "Why are there landless peasants?" is the key question for deterring the destruction of rain forests, and also the foundational question for the politics of social justice, then the movement to save rain forests needs to be closely linked with, if not virtually the same form, as the movement for social justice.

---

[1]Vandermeer, 1995.

[2]Roscha, 1905.

[3]Meyers, 1985.

[4]The similarity between the terminology here and modern literary criticism is not accidental. Much of our analysis was at least partially stimulated by Eagleton's *Literary Theory*, 1983.

[5]The basic idea is that "fitness," which is the ability to produce successful offspring (i.e. those that themselves reproduce offspring), is the "altitude" on a map, the coordinates of which are two traits of interest. The classical example is a skunk with two characteristics: fear and odor. A highly odoriferous skunk need not fear much, while a fearful skunk need not produce such a bad odor. Thus arise the alternative adaptive peaks of smelly and fearless, versus odorless and fearful, surrounded by valleys of smelly and fearful, or odorless and fearful.

[6]This metaphor is due to Lewontin (personal communication).

[7]We have here actually extended the notion of environmental landscape as trampoline, to include the human characteristic of social construction. Lewontin's original meaning did not take this extreme.

10

◆

Past Causes, Future Models, Present Action

The theme of this book has been, in various guises, the causes and consequences of rain forest destruction. Implicit in this theme has been the sub-theme that an alternative model of socioeconomic organization could drastically change those causes. Furthermore, certain programmatic conclusions seem to emerge from this analysis. In short, the book has sought to pose and answer three fundamental questions. 1) Question: What causes rain forest destruction? Answer: There is a web of causality, no single component of which is truly the cause. 2) Question: What is a model for the future? Answer: A planned mosaic, based on ecological and egalitarian principles. 3) Question: What is the political action plan? Answer: Intensify the struggle for social justice.

The Web of Causality

The notion of the *web of causality* is key to understanding why we face the problem of rain forest destruction in the first place. That has been a theme throughout this book. We illustrate the basic features of that web in figure 10.1. Beginning at the bottom of the figure we see that damaged rain forests will recuperate if not further damaged, but recuperate far more slowly if further modified by agriculture. The damaged rain forests themselves are created by either logging or modern agriculture, and further cleared by peasant farmers. But the latter's activities are a consequence of the opportunities created by logging as well as the ups and downs of the international market that causes modern agriculture to hire and fire workers. Modern agriculture needs those workers, as well as the land that it buys or steals from peasant farmers. Viewed as a web of causality, it is quite pointless to try to identify a single entity as the "true" cause. The true cause is the web itself.

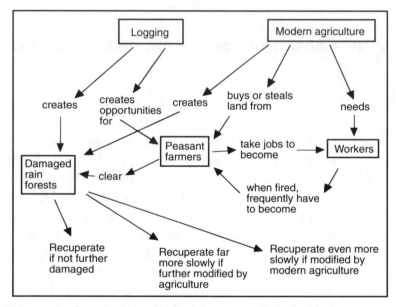

Figure 10.1. The web of causality for deforestation of rain forests.

Yet even this is an oversimplified picture. The web of causality is not just the farmers, loggers, modern agriculture, and workers. That is just a subweb embedded in a larger web, as illustrated in figure 10.2. With the expanded web, the international nature of the problem becomes apparent, including the key roles of the international banking system, national governments, the United States and other Developed World governments, as well as consumers and investors in the Developed World. This is the true web of causality, and it is a complicated interconnected one. Tweaking one strand is not likely to bring the whole thing down. Fighting a concerted battle to restructure the entire nature of the web is the only alternative.

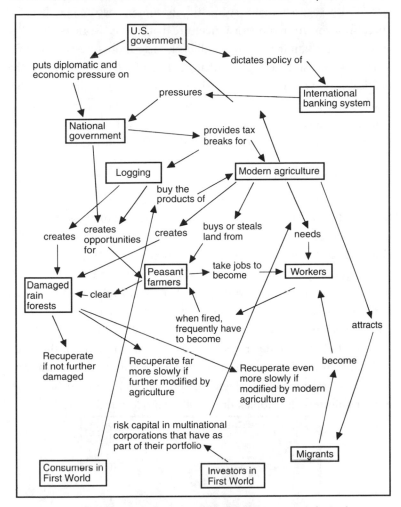

Figure 10.2. The expanded web of causality for deforestation of rain forests.

Furthermore, seeing the entire web of causality enables highly-focused political action to see its relation to other actions, and enables an analysis of consequences that may be dramatic, even though quite indirect. For example, organizing consumer boycotts can be seen as clearly attacking the arrow that connects the consumers to modern agriculture through the arrow "buys the products of." But, following through the logic of the web, it also suggests that a successful consumer boycott may likewise affect the arrow from modern agriculture that "needs" workers, thus creating more peasant farmers who will likely clear more forest. If a careful analysis of this situation reveals that the loss of jobs will be severe, the political-action agenda might then be expanded to form alliances with a local farm worker union calling for job security or a political movement that seeks secure land ownership for the increased number of peasant farmers that will surely be created if the boycott is a success.

The Planned Mosaic

In thinking of a model for the future, we are struck with two images from Central America. In figure 10.3 is a photograph of a poster that was on the wall of CORBANA, the technical research organization of the banana industry in Costa Rica. It is presented as a utopian future in which agricultural fields will be virtually reconstructed from their original state, crisscrossed with irrigation channels and subtended by plastic drainage tubes, all automatically installed with sophisticated laser-guided machinery. This is the vision of the modern technocrat. Nature has been fully tamed, down to the physical reconstruction of the water table under the soil. And in the model the isolated islands of pristine rain forest are surrounded by biological deserts of pesticide-drenched modern agriculture—the utopian dream of the modern technocrat.

Figure 10.3. Poster on the wall of CORBANA, the technical research institute of the banana companies in Costa Rica, emphasizing the technical mastery of nature.

The other image is presented in figure 10.4. This is a photograph of a poster that was hanging in the office of the Ministry of Agriculture of Nicaragua in 1988, during the Sandinista administration. It illustrates a mosaic of land uses, ranging from protected forests, to managed forests, to plantations, to sustainable agriculture. In short, it represents a diverse mosaic, where decisions about land use are tied to the capabilities of the land and the needs of people, not to the requirements of profit or repayment of past accumulated debt. This is quite a different vision. Nature has not been tamed, but is something of a partner with the humans who live there. This vision is the one that has a far greater chance of reordering that web of causality and stopping, perhaps even reversing, the tide of deforestation.

Figure 10.4. Poster on the wall of the Ministry of Agriculture in Nicaragua in 1988, emphasizing the ecological mosaic of land uses.

The Political Action Plan

All this is meaningless without a program of political action. While we never intended this book to be specifically about political action, certain principles do emerge from our analysis of the problems. Political action must focus on the web of causality and eschew single-issue foci. Calls for boycotts of tropical timbers or bananas need to be coupled with actions to change investment patterns and international banking pressures. Above all, political action plans must be formulated at least so they do not make the situation worse, which is certainly conceivable given the complex nature of the web of causality. But, as the various strands of this web are brought into the analysis, it appears obvious that political action needs to be focused, not only on rain forests and the subjects traditionally associated with them, but also on the same sorts of issues that motivated progressive forces in the past — issues of social justice. The same peasant farmers that formed the backbone of the Vietnamese liberation forces or the Salvadoran guerrillas, are the ones that are forced into the marginal existence that compels them to continually move into the forests. So the same issues that compelled progressive organizers in the past to form solidarity committees and anti-war protests are the issues that must be addressed if the destruction of rain forests is to be stopped.

Furthermore, just as the most effective political action in the past was organized in conjunction with, and to some extent under, the leadership of the people for whom social justice was being sought, so today political action should be coordinated with those same people. As that coordination proceeds, the alliances that form will inevitably lead to the reformulation of goals, which the rain forest conservation activist must acknowledge and be willing to respond to. The local people, quite obviously, must recognize something about the rain forest that is in their best interest to preserve, and it is the job of the progressive organizer to construct the political action so that such value is evident. In short, the alliance between the people that live in and around the rain forest and those from the outside that seek to stop the tide of rain forest destruction, must be a two-way alliance. If the people who live around the

Lacandon Forest in Mexico, for example, have as their major goal the reformulation of the Mexican political system, the rain forest conservationist must join the political movement to change that system—something that many would see as distant from the original goal of preserving rain forests. Political action to preserve rain forests, under the framework of the web of causality, will inevitably involve the serious activist in social justice issues that initially seem only marginally associated with the problem of rain forest destruction.

The web of causality, the landscape mosaic, and the social justice, political-action framework thus form the tripod upon which we seek to base the conclusions of this book. In the end we recall our initial image of slicing bananas on our cereal in the morning fading into the logger's saw slicing the trunks of the rain forest. We intend for this imagery to continually recall the interrelationship among various activities in the web of causality. We also intend to fix the image of banana workers laboring in substandard conditions before getting fired and being forced to seek land they can farm, sometimes converting a patch of rain forest for that purpose. A breakfast with sliced bananas thus becomes a metaphorical breakfast of biodiversity. And as integral parts of the same web of causality, slicing the bananas and cutting the trees are, in the end, more than just metaphor.

References

Adis, J., Y. D. Lubin, and G. G. Montgomery. 1984. Arthropods from the canopy of inundated and Terra firme forests near Manaus, Brazil, with critical considerations on the pyrethrum-fogging technique. *Studies on Neotropical Fauna and Environment* 19: 223–236.

Altieri, M. A. and S. B. Hecht. 1990. *Agroecology and Small Farm Development.* Macmillan, New York.

Altieri, M. A., M. K. Anderson, and L. C. Merrick. 1987. Peasant agriculture and the conservation of crop and wild plant resources. *Conservation Biology* 1: 49–58.

Andow, D. A. 1991. Vegetational diversity and arthropod population response. *Ann Rev. Entomol.* 36: 561–586.

Barry, T. 1987. *Roots of Rebellion: Land and Hunger in Central America.* South End Press, Boston.

Birdsey, R. A. and P. L.Weaver. 1982. The forest resources of Puerto Rico. USDA Forest Service. Southern Forest Experiment Station. *Resource Bulletin* SO-85 October, 1982.

Blaikie, P. and H. Brookfield, eds. 1987. *Land Degradation and Society.* Methuen, London.

Bonner, R. 1993. *At the Hands of Man: Perils and Hope for Africa's Wildlife.* Alfred A. Knopf, New York.

Boserup, E. 1965. *The Conditions of Agricultural Growth.* Aldine, New York.

Boucher, D., J. H. Vandermeer, M. A. Mallona, N. Zamora, and I. Perfecto. 1994. Resistance and resilience in a directly regenerating rainforest: Nicaraguan trees of the Vochysiaceae after Hurricane Joan. *J. of Forest Ecology.* 68: 127–136.

Boza, M. A. 1993. Conservation in action: Past, present, and future of the National Park System of Costa Rica. *Conservation Biology* 7: 239–247.

Brown, H. C. P. and V. G. Thomas. 1990. Ecological considerations for the future of food security in Africa. In Edwards, R., R. Lal, P. Madden, R. H. Miller, and G. House, eds., *Sustainable Agricultural Systems.* Soil and Water Conservation Society, Ankeny.

Brownrigg, L. A. 1985. *Home Gardening in International Development. What the Literature Shows.* The League for International Food Education. U. S. Agency for International Development, Washington, D.C.

Butterfield, 1994. The Regional Context: Land Colonization and Conservation in Sarapiquí. In McDade, L. A., K. S. Bawa, H. A. Hespenheide, and G. S. Hartshorn. (eds.) *La Selva: Ecology and Natural History of a Neotropical Rain Forest* 299–316. Univ. of Chicago Press, Chicago.

Butterfield, R. 1990. Native species for reforestation and land restoration: a case study from Costa Rica 3–14. Proceedings IUFRO World Congress. Montreal.

Chacón, J. C. and S. R. Gliessman. 1982. Use of the "non-weed" concept in traditional tropical agroecosystems of southeastern Mexico. *Agroecosystems* 8: 1–11.

Chapin, M. 1988. The seduction of models: Chinampa agriculture in Mexico. *Grassroots Development* 12: 8–17.

de Janvry, A. 1982. *The Agrarian Question and Reformism in Latin America.* Johns Hopkins Univ. Press, Baltimore.

Denslow, J. S. 1987. Tropical rainforest gaps and tree species diversity. *Ann. Rev. Ecol. Syst.* 18: 431–451.

Dover, M. and L. Talbot. 1987. *To Feed the Earth: Agroecology for Sustainable Development.* World Resources Institute, Washington, D. C.

Eagleton, T. 1983. *Literary Theory: An Introduction.* Univ. of Minnesota Press, Minneapolis.

Edelman, M. and J. Kenen (eds.) 1989. *The Costa Rican Reader.* Grove Press, New York.

Erwin, T. L. and J. C. Scott. 1980. Seasonal and size patterns, trophic structure, and richness of Coleoptera in the tropical arboreal ecosystem: the fauna of the tree *Luehea seemannii* Triana and Planch in the canal zone of Panama. *The Coleopterists Bulletin* 34: 305–322.

Espinoza, M. & R. Butterfield. 1989. Adaptabilidad de 13 especies nativas maderables bajo condiciones de plantación en las tierras bajas húmedas del Atlántico de Costa Rica. In R. Salazar (ed.) *Manejo y aprovechamientode plantaciones forestales con especies de uso múltiple.* Actas Reunión IUFRO, Guatemala, 1989, 159–172. CATIE. Turrialba, Costa Rica.

Gliessman, S. R. 1993. Managing diversity in traditional agroecosystems of tropical Mexico, 65–74. In Potter, C. S., J. I. Cohen, and D. Janczewski (eds.) *Perspectives on Biodiversity: Case Studies on Genetic Resource Conservation and Development.* Am. Association for the Advancement of Science Press, Washington, D. C.

Gómez-Poma, A. J. S. Flores and V. Sosa. 1987. The *"Pet Kot": A man-made tropical forest of the Maya.* Interciencia 12: 10–15.

Haberler, G. 1958. *Prosperity and Depression.* Harvard Univ. Press, Cambridge.

Hardin, G. 1968. The tragedy of the commons. *Science* 162: 1243–1248.

Hartshorn, G. S. 1987. Application of gap theory to tropical forest management: natural regeneration on strip clear-cuts in the Peruvian Amazon. *Ecology* 70: 567–569.

Hölldobler, B. and E. O. Wilson. 1990. *The Ants.* Harvard Univ. Press, Cambridge.

Huss, J. and M. Sutisna. Conversion of exploited natural Dipterocarp forests into semi-natural production forests. In H. Lieth & M. Lohmann. (eds.) *Restoration of Tropical Forest Ecosystems.* Proc. Symp., Oct. 1991, 145–153. Kluwer Academic Publ. Dordrecht, Netherlands.

Huston, M. 1993. Biological diversity. Soils and economics. *Science* 262: 1676–1680.

Huston, M. A. 1995. *Biological Diversity: The Coexistence of Species in Changing Landscapes.* Cambridge Univ. Press, Cambridge.

Immerman, R. H. 1982. *The CIA in Guatemala: The Foreign Policy of Intervention.* Univ. of Texas Press, Austin.

Janzen, D. H. and T. W. Schoener. 1968. Differences in insect abundance and diversity between wetter and drier sites during a tropical dry season. *Ecology* 49: 96–110.

Johns, A. D. 1985. Selective logging and wildlife conservation in tropical rain forest: problems and recommendations. *Biological Conservation* 31: 355–375.

Johns, A. D. 1991. Responses of Amazonian rain forest birds to habitat modification. *J. of Tropical Ecology* 76: 417–437.

Kaimowitz,, D. 1986. Nicaragua's Agrarian Reform: Six Years Later. In Rosset, P. and J. Vandermeer, (eds.). The *Nicaraguan Reader: Documents of a Revolution Under Fire,* Grove Press, New York.

Kanth, R. (ed.) 1994. *Paradigms in Economic Development.* M. E. Sharpe, Armonk, New York.

Karr, J. R. 1971. Structure of avian communities in selected Panama and Illinois habitats. *Ecological Monographs* 41: 207–233.

Lewis, S. A. 1992. Banana Bonanza: Multinational Fruit Companies in Costa Rica. *The Ecologist,* 22: 289–290.

Lewontin, R. 1982. Agricultural research and the penetration of capital. *Science for the People* 14: 12–17.

Lohmann, L. 1990. Commercial tree plantations in Thailand: Deforestation by any other name. *The Ecologist* 20(1): 9–17.

Lovejoy, T. E., J. M. Rankin, R. D. J. Bierregaard, K. S. Brown, Jr., L. H. Emmons, and M. E. Van der Voort. 1984. Ecosystem decay in Amazon forest remnants. In Niteki, M. H. (ed.) *Extinctions,* 295–321. Univ. of Chicago Press, Chicago.

Lovejoy, T. E., R. D. Bierregaard, Jr. A. B. Rylands, J. R. Malcolm, C. F. Quintela, L. H. Harper, K. S. Brown, Jr., A. H. Powell, G. V. N. Powell, H. O. R. Schubart, and M. B. Hays. 1986. Edges and other effects of isolation on Amazon forest fragments. In Solé, M. E. (ed.) *Conservation Biology: The Science of Scarcity and Diversity,* 257–285. Sinauer, Sunderland.

Lugo, A. 1988. Estimating reductions in the diversity of tropical forest species. In E. O. Wilson (ed.), *Biodiversity.* National Academy Press, Washington, D.C.

Lugo, A. E. 1992. Comparison of tropical tree plantations with secondary forest of similar age. *Ecological Monographs* 62(1): 1–41.

Mabberley, D. J. 1992. *Tropical Rain Forest Ecology.* Chapman and Hall, New York.

Maury-Lechon, G. 1993. Biological characters and plasticity of juvenile tree stages to restore degraded tropical forests. In H. Lieth & M. Lohmann. *Restoration of Tropical Forest Ecosystems,* 37–46. Proc. Symp., Oct. 1991. Kluwer Academic Publ. Dordrecht, Netherlands.

McWilliams, C. 1939. *Factories in the Field: The Story of Migratory Farm Labor in California.* Little Brown & Co., Boston.

McCann, T. P. 1976. *An American Company: The Tragedy of United Fruit.* Ed. Henry Scammel, New York.

Melville, T. and M. Melville. 1971. *Guatemala: The Politics of Land Ownership,* New York.

Méndez, C. 1989. *Fight for the Forest: Chico Méndez in His Own Words.* Latin America Bureau, London.

Michael, E. D. and P. I. Thornburgh. 1971. Immediate effects of hardwood removal and prescribed burning on bird populations. *Southwest Naturalist* 15: 359–370.

Michon, G. 1983. Village-forest gardens in West Java. In Huxley, P. A. (ed.) *Plant Research and Agroforestry,* 13–24. International Council for Research in Agroforestry (ICRAF), Nairobi.

Minc, L. D. and J. H. Vandermeer. 1990. The origin and spread of agriculture. In Carroll, C. R., J. H. Vandermeer, and P. M. Rosset, *Agroecology,* 65–145. McGraw Hill Pub. Co, New York.

Montangnini, F. 1994. Agricultural systems in the La Selva region. In McDade, L. A., K. S. Bawa, H. A. Hespenheide, and G. S. Hartshorn. (eds.) *La Selva: Ecology and Natural History of a Neotropical Rain Forest,* 307–316. Univ. of Chicago Press, Chicago

Mooney, Harold A. and M. Gordon (eds.) 1983. *Disturbance and Ecosystems.* Springer-Verlag, Berlin

Morgan, D. 1979. *Merchants of Grain.* Viking, New York.

Myers, N. 1980. *The Conservation of Tropical Moist Forests.* Nat. Acad. Sci., Washington, D.C.

Myers, N. 1985. *The Primary Source: Tropical Forests and Our Future.* W. W. Norton and Co., New York and London.

National Research Council. 1993. *Sustainable Agriculture and the Environment in the Humid Tropics.* National Academy Press, Washington, D.C.

Nietchmann, B. 1973. *Between Land and Water: The Subsistence Ecology of the Miskito Indians, Eastern Nicaragua.* Seminar Press, New York.

Paoletti, M. G. 1988. Soil invertebrates in cultivated and uncultivated soils in northern Italy. *Firenze* 71: 501–563.

Perfecto, I. and J. H. Vandermeer. 1994. The ant fauna of a transforming agroecosystem in Central America. *Trends in Agricultural Science* 2: 7–13

Perfecto, I. and R. Snelling. 1995. Ant diversity in the coffee agroecosystem in Costa Rica. *Ecological Applications,* in press.

Perfecto, I., M. A. Mallona, I. Granzow de la Cerda, and J. H. Vandermeer. 1994. Los recursos terrestres de la Costa Caribeña de Nicaragua: Hacia una filosofia de sostenibilidad. *Wani,* 15: 46–59 .

Phillips, O. L. and A. H. Gentry. 1994. *Science* 263: 954.

Pickett, S. T. A. and P. S. White. 1985. *The Ecology of Natural Disturbance and Patch Dynamics.* Academic Press, New York.

Pimentel, D., V. Stachow, D. A. Takacs, H. W. Burbaker, A. R. Dumas, J. H. Meaney, J. A. S. O'Neal, D. E. Onsi, and D. B. Corzinus. 1992. Conserving biological diversity in agricultural/forestry systems. *Bioscience* 42:

354–362.

Poore, D. 1989. *No Timber without Trees*. Earthscan Publications Ltd., London.

Rabe, S. G. 1988. *Eisenhower and Latin America: The Foreign Policy of Anticommunism*. Univ. of North Carolina Press, Chapel Hill.

Redclift, M. 1987. 'Raised bed' agriculture in pre-Columbian Central and South America: A traditional solution to the problem of 'sustainable' farming systems. *Biol. Agric. Hort. Int. J.* 5: 51–59.

Richter, D. D. and L. I. Babbar. 1991. Soil diversity in the tropics. *Advances in Ecological Res.* 321: 315–389.

Ricklef, R. E. and D. Schluter. 1993. *Species Diversity in Ecological Communities: Historical and Geographical Perspectives*. Univ. of Chicago Press, Chicago.

Risch, S. J., D. Andow, and M. Altieri. 1983. Agroecosystem diversity and pest control: data, tentative conclusions, and new research directions. *Environ. Entomol.* 12: 625–629.

Rocha, J. 1905. *Memorandum de un viaje* Editorial El Mercurio, Bogota.

Rosset, P. and M. Benjamin. 1994. *The Greening of Cuba*. Ocean Press.

Rosset, P. M. and J. H. Vandermeer (eds.) 1986. *Nicaragua, Unfinished Revolution: The New Nicaraguan Reader*. Grove Press, New York.

Russell, E. 1993. War on insects: Warfare, insecticides, and environmental change in the United States, 1970–1945. PhD thesis, Univ. Michigan.

Samuelson, P. A. 1970. *Economics*. McGraw-Hill, New York.

Sanchez, P. A. 1976. *Properties and management of soils in the tropics*. John Wiley and Sons, New York.

Schoener, T. W. 1993. On the relative importances of direct versus indirect effects in ecological communities. In H. Kawanabe, J. E. Cohen, and K. Iwasaki, (eds.) *Mutualism and Community Organization: Behavioral, Theoretical and Food Web Approaches*, 365–411. Oxford Univ. Press, Oxford.

Seligson, M. A. 1980. *Peasants of Costa Rica and the Development of Agrarian Capitalism*. Univ. of Wisconsin Press, Madison.

Shiva, V. 1993. *Monocultures of the Mind*. Zed Books Ltd., London.

Soemarwoto, O., I. Soemarwoto, Karyano, E. M. Soekartadiredja, and A. Famlan. 1985. The Javanese home gardens as an integrated ecosystem. *Food Nutrition Bull.* 7: 44–47.

Sousa, W. P. 1984. The role of disturbance in natural communities. *Ann. Rev. Ecol. Syst.* 15: 353–391.

Stoler, A. L. 1978. Garden use and household economy in rural Java. *Bull. Indonesian Economic Studies* 14: 85–101.

Swift, M. J., O. W. Heal, and J. M. Anderson. 1979. Decomposition in terrestrial ecosystems. *Studies in Ecology*, Vol 5. Blackwell Scientific, Oxford.

Swift, M., J. H. Vandermeer, J. Anderson, R. Ramakrishnan, B. Hastings, and C. Ong. 1994. Biodiversity changes in agricultural transformation. In Mooney, H. ed., *Global change and biodiversity*, in press.

174 Breakfast of Biodiversity

Tangley, L. 1990. Cataloging Costa Rica's diversity. *Bioscience* 40. 633–636.

Terborgh, J. 1992. *Diversity and the Tropical Rain Forest.* Scientific American Library. New York.

Thrupp, L. A. 1990. Environmental initiatives in Costa Rica: A political ecology perspective. *Society and Natural Resources* 33: 243–256.

Thrupp, L. A. 1991. The human guinea pigs of Rio Frio: Standard Fruit keeps its eye on the bottom line. *The Progressive,* April 1991: 28–30.

USDA (United States Department of Agriculture). 1975. *Soil Taxonomy.* Agriculture Handbook #436. Government Printing Office, Washington, D.C.

Vandermeer, J. H. 1991. The political economy of sustainable development: The Southern Atlantic Coast of Nicaragua. *The Centennial Review,* XXXV: 265–294.

Vandermeer, J. H. 1995. *Reconstructing Biology: Genetics and Ecology in the New World Order.* John Wiley and Sons, New York.

Vandermeer, J. H., D. Boucher, and I. Perfecto. 1991. Conservation in Nicaragua and Costa Rica: Indirect consequences of social policy. *INTECOL Bulletin,* January.

Vandermeer, J. H., M. A. Mallona, D. Boucher, K. Yih, and I. Perfecto. 1995. Three years of ingrowth following catastrophic hurricane damage on the Caribbean coast of Nicaragua: evidence in support of the direct regeneration hypothesis. *J. of Tropical Ecology,* in press.

Wallerstein, I. 1980. *The Modern World System II: Mercantilism and the Consolidation of the European World-Economy, 1260–1750.* Academic Press, New York.

Webb, W. L., D. F. Behrend, and B. Saisoin. 1977. Effect of logging on songbird populations in a northern hardwood forest. *Wildlife Monographs* 55.

Werner, E. E. 1992. Individual behavior and higher-order species interactions. *Amer. Naturalist* 140: S5–S32.

Westoby, J. 1989. *Introduction to World Forestry.* Basil Blackwell, Oxford.

Whitmore, T. C. 1991. *An Introduction to Tropical Rain Forests.* Claredon Press, Oxford.

Wilkin, G. 1987. *Good Farmers.* University of California Press, Berkeley.

Wilson, E. O. 1987. The arboreal ant fauna of Peruvian Amazon forests: A first assessment. *Biotropica* 19: 245–251.

Wilson, E. O. 1988. The current state of biological diversity. In Wilson, E. O. (ed.) *Biodiversity,* 3–8. National Academy Press, Washington, D.C.

World Resources Institute. 1990. *World Resources 90–91.* Oxford Univ. Press, Oxford.

World Resources Institute. 1994. *World Resources 94–95.* Oxford Univ. Press, Oxford.

Yih, K. , D. Boucher, J. H. Vandermeer, and N. Zamora. 1991. Recovery of the rainforest of southeastern Nicaragua after destruction by Hurricane Joan. *Biotropica* 23: 106–113.

Index

About the Authors

John Vandermeer is Alfred Thurneau Professor of Biology at the University of Michigan. He has been actively involved in research on the ecology of tropical rain forests for the past 25 years, concentrating his efforts in Nicaragua and Costa Rica. He is currently conducting research on the regeneration of Nicaraguan rain forests following the damage caused by Hurricane Joan in 1988. He has written or edited 6 books, including *The Nicaragua Reader, Agroecology,* and *Biology as a Social Weapon,* and over 100 articles.

Ivette Perfecto, a native of Puerto Rico, is Associate Professor in the School of Natural Resources and the Environment, University of Michigan. She has been involved in ecological research in rain forest areas for 10 years, concentrating her efforts in Nicaragua, Costa Rica, and Puerto Rico. She is currently examining the ecological consequences of agricultural modernization in the Tropics.

Vandana Shiva, a physicist and philosopher, is Director of the Research Foundation for Science, Technology and Natural Resource Policy, Dehradun, India and is the Science and Environment Advisor of the Third World Network. She is the author of *Staying Alive: Women, Ecology and Development; The Violence of the Green Revolution: Third World Agriculture, Ecology and Politics; Biodiversity: Social and Ecological Perspectives;* and *Ecofeminism.*

Food First Publications

Alternatives to the Peace Corps: A Directory of Third World and US Volunteer Opportunities by Becky Buell and Annette Olson. Now in its sixth edition, this guide provides essential information on voluntary service organizations, technical service programs, work brigades, study tours, as well as alternative travel in the Third World and offers options to the Peace Corps as the principle route for people wishing to gain international experience. $6.95

BASTA! Land and the Zapatista Rebellion in Chiapas by George Collier with Elizabeth Lowery Quaratiello. The authors examine the root causes of the Zapatista uprising in southern Mexico and outline the local, national, and international forces that created a situation ripe for a violent response. $12.95

Brave New Third World? Strategies for Survival in the Global Economy by Walden Bello. Can Third World countries finish the next decade as vibrant societies? Or will they be even more firmly in the grip of underdevelopment? The outcome, Bello argues, depends on their ability to adopt a program of democratic development which would place them on equal footing in the global economy. $6.00 (Development Report)

Chile's Free-Market Miracle: A Second Look by Joseph Collins and John Lear. The economic policies behind the boom experienced by Chile in the 1980s under Pinochet, and continuing today, are widely touted as a model for the Third World. The authors take a closer look at the Chilean experience and uncover the downside of the model: chronic poverty and environmental devastation. $15.95

Circle of Poison: Pesticides and People in a Hungry World by David Weir and Mark Shapiro. In the best investigative style, this popular exposé documents the global scandal of corporate and government exportation of pesticides and reveals the threat to the health of consumers and workers throughout the world. $7.95

Dark Victory: The U.S., Structural Adjustment, and Global Poverty by Walden Bello, with Shea Cunningham and Bill Rau. Offers an understanding of why poverty has deepened in many countries, and analyzes the impact of Reagan-Bush economic policies: a decline of living standards in much of the Third World and the U.S. The challenge for progressives in the 1990's is to articulate a new agenda because the people of the South and North suffer from the same process that preserves the interests of a global minority. $12.95

Dragons in Distress: Asia's Miracle Economies in Crisis by Walden Bello and Stephanie Rosenfeld. Economists often refer to South Korea, Taiwan, and Singapore as "miracle economies," and technocrats regard them as models for the rest of the Third World. The authors challenge these established notions and show how, after three decades of rapid growth, these economies are entering a period of crisis. The authors offer policy recommendations for structural change to break the NICs unhealthy dependence on Japan and the U.S., and they critically examine both the positive and negative lessons of the NIC experience for the Third World. $12.95

Education for Action: Graduate Studies with a Focus on Social Change by Andrea Freedman and Nooshi Borhan. A guide to progressive graduate programs and educators in agriculture, anthropology, development studies, economics, ethnic studies, history, law, management, peace studies, political science, public health, sociology, urban planning and women's studies. $6.95

Kerala: Radical Reform as Development in an Indian State by R.W. Franke and B.H. Chasin. Analyzes both the achievements and the limitations of the Kerala experience. In the last eighty years, the Indian state of Kerala has undergone an experiment in the use of radical reform as a development strategy that has brought it some of the Third World's highest levels of health, education, and social justice. 1994 revised edition $9.95

Needless Hunger: Voices from a Bangladesh Village by James Boyce and Betsy Hartmann. The global analysis of Food First is vividly captured here in a single village. The root causes of hunger emerge through the stories of both village landowners and peasants who live at the margin of survival. Now in its sixth printing! $6.95

No Free Lunch: Food and Revolution in Cuba Today by Medea Benjamin, Joseph Collins and Michael Scott. Based on sources not readily available to Western researchers, this examination of Cuba's food and farming system confirms that Cuba is the only Latin American country to have eradicated hunger. $9.95

People and Power in the Pacific: The Struggle for the Post-Cold War Order by Walden Bello. Examines the extent to which events in the Asia-Pacific region reflect the so-called new world order; the future role of the U.S. and the emergence of Japan as a key economic power on the world stage. $12.00

The Philippines: Fire on the Rim by Joseph Collins. Looks at the realities following the People Power revolution in the Philippines. A choir of voices from peasants, plantation managers, clergy, farmers, prostitutes who serve U.S. military bases, mercenaries, revolutionaries and others, speak out. *Hardcover* $9.50, *Paper* $5.00.

Taking Population Seriously by Frances Moore Lappé and Rachel Schurman. The authors conclude that high fertility is a response to antidemocratic power structures that leave people with little choice but to have many children. The authors do not see the solution as more repressive population control, but instead argue for education and improved standard of living. $7.95

Trading Freedom: How Free Trade Affects Our Lives, Work, and Environment edited by John Cavanagh, John Gershman, Karen Baker and Gretchen Helmke. Contributors from Mexico, Canada and the U.S. analyze the North American Free Trade Agreement. Drawing on the experiences of communities in Canada, the U.S., and Mexico, this comprehensive collection provides a hard-hitting critique of the current proposals for a continental free trade zone through an intensive examination of its impact on the environment, workers, consumers, and women. $5.00

Curricula

Exploding the Hunger Myths: A High School Curriculum by Sonja Williams. With an emphasis on hunger, twenty-five activities provide a variety of positive discovery experiences—role playing, simulation, interviewing, writing, drawing—to help students understand the real underlying causes of hunger and how problems they thought were inevitable can be changed. 200 pages, 8.5x11 with charts, reproducible illustrated hand-outs, resource guide and glossary. $15.00

Food First Curriculum by Laurie Rubin. Six delightfully illustrated units span a range of compelling topics including the path of food from farm to table, why people in other parts of the world do things differently, and how young people can help make changes in their communities. 146 pages, three-hole punched, 8.5x11 with worksheets and teacher's resources. $12.00

Food First Comic by Leonard Rifas. An inquisitive teenager sets out to discover the roots of hunger. Her quest is illustrated with wit and imagination by Rifas, who has based his comic on *World Hunger: Twelve Myths.* $1.00

Write or call our distributor to place book orders. All orders must be prepaid. Please add $4.00 for the first book and $1.00 for each additional book for shipping and handling.

Subterranean Company
Box 160, 265 South 5th Street
Monroe, OR 97456
(800) 274-7826

About Food First

Food First, publisher of this book, is a nonprofit research and education-for-action center. We work to identify the root causes of hunger and poverty in the United States and around the world, and to educate the public as well as policy-makers about these problems.

The world has never produced so much food as it does today—more than enough to feed every child, woman, and man. Yet hunger is on the rise, with more than one billion people around the world going without enough to eat.

Food First research has demonstrated that hunger and poverty are not inevitable. Our publications reveal how scarcity and overpopulation, long believed to be the causes of hunger, are instead symptoms—symptoms of an ever-increasing concentration of control over food-producing resources in the hands of a few, depriving so many people of the power to feed themselves. In 55 countries and 20 languages, Food First materials and activism are freeing people from the grip of despair and laying the groundwork—in ideas and action—for a more democratically controlled food system that will meet the needs of all.

An Invitation to Join Us

Private contributions and membership dues form Food First's financial base. Because we are not tied to any government, corporation, or university, we can speak with strong independent voices, free of ideological formulas. The success of our programs depends not only on dedicated volunteers and staff, but on financial activists as well. All our efforts toward ending hunger are made possible by membership dues or gifts from individuals, small foundations, and religious organizations. We accept no government or corporate funding.

Each new and continuing member strengthens our effort to change a hungry world. We'd like to invite you to join in this effort. As a member you will receive a 20 percent discount on all Food First books. You will also receive our quarterly publication, Food First News and Views, and our timely Backgrounders which provide information and suggestions for action on current food and hunger crises in the United States and around the world.

All contributions to Food First are tax deductible.

To join us in putting food first, just clip and return the attached coupon to:

Food First/Institute for Food and Development Policy,
398 60th Street, Oakland, CA 94618
(510) 654-4400

Research internship opportunities are also available.
Call or write us for more information.

Name _____

Address _____

City/State/Zip _____

Daytime Phone (   ) _____

❏ I want to join Food First and
receive a 20% discount on this and
all subsequent orders. Enclosed
is my tax-deductible contribution of:
☐ $100   ☐ $50   ☐ $30

Page	Item Description	Qty	Unit Cost	Total
T-shirts ☐ XL ☐ L ☐ M ☐ S			$12.00	

Payment Method: ☐ Check   ☐ Money Order   ☐ Mastercard   ☐ Visa

For gift mailings, please see other side of this coupon.

Name on Card _____

Card Number _____Exp. Date _____

Signature _____

Member discount -20%	$ _____
CA Residents 8.5%	$ _____
SUBTOTAL	$ _____
Postage/15%-UPS/20% ($2 min.)	$ _____
Membership(s)	$ _____
Contribution	$ _____
TOTAL ENCLOSED	$ _____

Make check payable to Subterranean Company, Box 160, 265 South 5th St., Monroe, OR 97456

- -

Please send a Gift Membership to:

Name _____

Address _____

City/State/Zip _____

From _____

Please send a Gift Book to:

Name _____

Address _____

City/State/Zip _____

From _____

Please send a Resource Catalog to:

Name _____

Address _____

City/State/Zip _____

Name _____

Address _____

City/State/Zip _____

Name _____

Address _____

City/State/Zip _____

Name _____

Address _____

City/State/Zip _____